Plumbing Systems
Analysis, Design, and Construction

TIM WENTZ

University of Nebraska at Lincoln

Prentice Hall
Upper Saddle River, New Jersey   Columbus, Ohio

Library of Congress Cataloging-in-Publication Data
Wentz, Tim.
　　Plumbing systems : analysis, design, and construction / Tim Wentz.
　　　　p.　cm.
　　Includes index.
　　ISBN 0-13-235284-2
　　1. Plumbing.　I. Title.
TH6123.W46　1997
696'.1—dc20　　　　　　　　　　　　　　　　　　　　　　　　　96-20096
　　　　　　　　　　　　　　　　　　　　　　　　　　　　　　　　CIP

Editor: Ed Francis
Production Editor: Rex Davidson
Design Coordinator: Julia Zonneveld Van Hook
Text Designer: STELLARViSIONs
Cover Designer: Brian Deep
Production Manager: Laura Messerly
Marketing Manager: Danny Hoyt
Electronic Text Management: Marilyn Wilson Phelps, Matthew Williams, Karen L. Bretz, Tracey Ward
Illustrations: Christine Haggerty

This book was set in Dutch 823 and Swiss 821 by Prentice Hall and was printed and bound by Quebecor Printing/Book Press. The cover was printed by Phoenix Color Corp.

© 1997 by Prentice-Hall, Inc.
Simon & Schuster/A Viacom Company
Upper Saddle River, New Jersey 07458

All rights reserved. No part of this book may be reproduced, in any form or by any means, without permission in writing from the publisher.

Printed in the United States of America

10　9　8　7　6　5　4　3　2　1

ISBN: 0-13-235284-2

Prentice-Hall International (UK) Limited, *London*
Prentice-Hall of Australia Pty. Limited, *Sydney*
Prentice-Hall of Canada, Inc., *Toronto*
Prentice-Hall Hispanoamericana, S. A., *Mexico*
Prentice-Hall of India Private Limited, *New Delhi*
Prentice-Hall of Japan, Inc., *Tokyo*
Simon & Schuster Asia Pte. Ltd., *Singapore*
Editora Prentice-Hall do Brasil, Ltda., *Rio de Janeiro*

Preface

A construction project is a diverse and complex undertaking that requires a coordinated effort from specialists in a number of different disciplines. Mechanical engineers, structural engineers, electrical engineers, and others work together to translate the owner's wishes into a building that meets the owner's needs. Construction managers, architects, and other managers need to have a working knowledge of each of these different disciplines if they are to participate effectively in management and problem-solving decisions.

Accordingly, this text was written specifically for construction managers, architects, and other managers who want to know and understand the fundamentals of the plumbing system. The text focuses on how and why plumbing systems work and how plumbing systems relate to other elements of the construction. As such, most of the material is presented in a "systems" approach to plumbing, as opposed to trying to teach the design engineering aspects of plumbing. However, to make the system easier to understand, some design fundamentals are presented and explained.

This text is based, in part, on the new ANSI A40-1993 Standard, *Safety Requirements for Plumbing*. This Standard was recently published by the joint task force of the Mechanical Contractors Association of America (MCAA) and the National Association of Plumbing, Heating, Cooling Contractors (PHCC). Since its release, the International Association of Plumbing and Mechanical Officials (IAPMO) have also co-sponsored the A40 Standard.

The A40 Standard is one of the first efforts to produce a single national code or standard. This is highly significant, because the field of plumbing codes and standards has long been fragmented. For example, the standard plumbing codes currently in use throughout the United States are the Uniform Plumbing Code (UPC), the National Standard Plumbing Code, the Southern Plumbing Code, and the BOCA (Building Officials and Code Administrators) Code. Each is widely used throughout the United States, and each code, although similar, has significant differences.

If that is not confusing enough, keep in mind that each locality usually adopts a standard code with alternates or addenda that, in effect, create another unique code. In short, there are hundreds, if not thousands, of plumbing codes in use in construction. For this reason, you will note that the text contains the terms "usually," "typically," and "often" more than one would like in a text. Hopefully, this should not pose any problems as we discuss and explore the fundamentals of plumbing systems. It is imperative, however, that you familiarize yourself as a construction manager with the local plumbing code that has jurisdiction over your specific project.

Acknowledgments

As with any undertaking of this type, there are a large number of people who deserve special recognition.

Special thanks goes to the National Association of Plumbing, Heating, Cooling Contractors (PHCC) and the Mechanical Contractors Association of America (MCAA) for allowing me to reproduce the tables and illustrations from the National Standard Plumbing Code and the ANSI A40-1993 code. Also, a special thank you goes out to the Plumbing and Drainage Institute (PDI) and Plumbing Engineer magazine for their invaluable assistance.

A number of manufacturers were kind enough to provide illustrations, tables, and charts for this text. The Kohler Company, Bell & Gossett, J. R. Smith Co., and Nibco Valve Co. all contributed greatly to the clarity of this text.

A number of individuals also deserve special thanks. Prominent among these are Mr. Cliff Dean of the Verne Simmonds Co., Mr. Bill Biggs of Lincoln Winnelson, and Mr. Bob Converse of Wentz Plumbing and Heating Co. These gentlemen not only assisted me on this project, but also have taught me a lot about plumbing systems on numerous projects over the years.

Additionally, I want to thank David Hanna of Ferris State University for reviewing the manuscript.

The list of thanks would not be complete without mentioning my two "mentors," Mr. Bob Cochran, the best engineer I ever met; and my father, Stan Wentz, the best contractor I ever met. From these two gentlemen, I have gained an invaluable education, a lot of which I have tried to include in this text.

Finally, I want to thank my wife, Marsha, who tolerates all my idiosyncracies and seems to have weathered the production of this text far better than myself.

Contents

1 Introduction   1

    The Intent of Plumbing Systems   1
    The Language of Plumbing   2
    Review Questions   2

2 Drain, Waste, and Vent Fundamentals   3

    Drainage Systems Flow by Gravity   3
    Every Plumbing Fixture Must Have a Trap   6
    Every Trap Must Have a Vent   12
    Review Questions   20

Materials 21

Selection Criteria   21
Material Types   22
    Plastics   23
    Cast Irons and Carbon Steels   25
    Nonferrous Metals   28
    Stone-based Piping   31
Drain, Waste, and Vent Fittings   32
Supports and Hangers   35
Valves   37
Review Questions   39

Sources of Water Supply and Points of Wastewater Disposal   41

Wastewater Disposal   41
    Septic Tank with Distribution Fields   42
    Aeration Systems   44
    Stabilization Ponds   45
Water Sources   46
    Public Water Supply Systems   46
    Private Water Sources   47
Review Questions   47

5
Plumbing Fixtures   49

Determining the Type of Plumbing Fixture   49
    Water Closets   50
    Urinals   52
    Sinks   53
    Bathtubs and Showers   54
    Drinking Fountains   56
    Floor Drains, Roof Drains, and Cleanouts   56
    Sewage Ejectors, Sump Pumps, and Booster Pumps   57
Determining the Quantity of Plumbing Fixtures   60
Deciding Where to Place the Plumbing Fixtures   60
Review Questions   62

Sizing the Drain, Waste, and Vent System   65

Drainage Fixture Units   65
Riser Diagrams   67
Sizing the Drain and Waste System   69
Sizing of the Vent System   73
Review Questions   76

7 Storm Water Systems — 79

Introduction 79
Fundamentals 80
Sizing Storm Drainage Systems 83
Review Questions 92

8 Domestic Water System — 93

Pressure Differential 93
Noise 95
Temperature 97
Public Health 100
Expansion and Contraction 105
Sizing the Domestic Water System 107
Sizing a Water Booster Pump 121
Domestic Water Heating Fundamentals 124
Tank-Type Water Heaters 126
Instantaneous Water Heaters 127
Multiple Temperature Applications 129
Sizing a Hot Water Heater 130
Review Questions 131

Appendices 133

Index 213

CHAPTER 1

Introduction

THE INTENT OF PLUMBING SYSTEMS

One of the first tools that a construction manager learns is how to *interpret* or read plans and specifications for a construction project. This process of interpretation begins with the design intent of the engineer and architect, as expressed in the contract documents. The same is true with plumbing systems. To properly interpret any plumbing system, you must first start with the intent of the system.

There are, literally, hundreds of adaptations of the four main plumbing codes used in the United States. The good news is that all codes have the same basic intent: to provide a safe, healthy environment through a "properly designed, acceptably installed and adequately maintained plumbing system."[1] Stated simply, a healthy environment provides for an adequate amount of safe, potable water, and a safe, healthy method of collecting and disposing of liquid and solid wastes.

This basic intent is described by the 20 Principles of Plumbing, found in Appendix A, reprinted from the ANSI-A40 Standard. Although these principles are very simple and straightforward, they provide a clear understanding of the intent of the plumbing system. If you are lost or confused on a question

---

[1] ANSI A40-1993 Safety Requirements for Plumbing.

involving the plumbing system (and we all are, from time to time), a quick review of these principles will often clear away the haze and allow you to reach a decision consistent with the intent of the Standard.

THE LANGUAGE OF PLUMBING

As with any other discipline, plumbing has its own language. As a construction manager, you do not want to sit at a high-powered meeting with the mechanical engineer, the mechanical contractor, and the owner and feel that all of the parties are speaking in a foreign language. To communicate and successfully participate in the problem-solving process, you need to be familiar with the terms and phrases used in the industry. You also need to be forewarned that sometimes the differences in terms is very subtle. For example, we will soon discover that there is a significant difference between a vent stack and a stack vent.

Appendix B contains an illustrated listing of the more common terms and phrases used in the plumbing industry. You should study these terms and phrases until you are familiar with the language of plumbing systems.

You should also be aware that the plumbing system is usually divided into two subsystems: the drain, waste, and vent system (usually abbreviated DWV); and the domestic water system (sometimes referred to as the potable water system). The DWV system is also subdivided into two different categories: the sanitary sewer system, which involves the collection and disposal of human wastes; and the storm sewer system, which involves the collection and disposal of rainwater, snow melt, and other clear water wastes. We will explore each of these main categories separately.

REVIEW QUESTIONS

1. What is the basic intent of any plumbing system?
2. What is a potable water system?
3. What is a DWV system?
4. Describe a plumbing system.
5. Select any two of the 20 Principles of Plumbing and describe how this principle protects the public safety.

CHAPTER

Drain, Waste, and Vent Fundamentals

The drain, waste, and vent system (DWV) collects the discharge of various plumbing fixtures within a building and disposes of the waste by directing it to a point of disposal, usually a city sewer main. The secret to understanding DWV systems is to understand first the fundamentals of how and why they work.

DRAINAGE SYSTEMS FLOW BY GRAVITY

DWV systems typically rely on gravity to pull the effluent along a piping system. This is sometimes referred to as the "First Great Truth of Plumbing" (i.e., it does flow downhill). Most other mechanical systems, such as the domestic water system we will study later, rely on a pressure differential to move a fluid.

It follows then that all DWV piping must be installed with a slope or grade. Logic also dictates that the slope of the waste pipe and the capacity of the waste pipe are related. In other words, the greater the slope of the pipe, the greater the carrying capacity of the pipe. The same relationship holds true for the velocity of the fluid. That is, the greater the slope, the higher the velocity of the fluid within the pipe. This point is illustrated in Table 2–1.

For example, Table 2–1 shows that a 4" drain line, at ⅛" of slope per foot of pipe, will carry 37.8 gallons per minute (gpm) of fluid with a velocity of

Table 2-1 Approximate discharge rates and velocities in sloping drains.

Flowing Half Full Discharge Rate and Velocity

Actual Inside Diameter of Pipe, in.	Slope 1/16 in./ft Discharge gal/min.	Velocity ft/sec	Slope 1/8 in./ft Discharge gal/min	Velocity ft/sec	Slope 1/4 in./ft Discharge gal/min	Velocity ft/sec	Slope 1/2 in./ft Discharge gal/min	Velocity ft/sec
1¼"							3.4	1.78
1⅜"					3.13	1.34	4.44	1.9
1½"					3.91	1.42	5.53	2.01
1⅝"					4.81	1.5	6.8	2.12
2"			10.8	1.41	8.42	1.72	11.9	2.43
2½"			17.6	1.59	15.3	1.99	21.6	2.82
3"	26.7	1.36	37.8	1.93	24.8	2.25	35.1	3.19
4"	48.3	1.58	68.3	2.23	53.4	2.73	75.5	3.86
5"	78.5	1.78	111	2.52	96.6	3.16	137	4.47
6"	170	2.17	240	3.07	157	3.57	222	5.04
8"	308	2.52	436	3.56	340	4.34	480	6.13
10"	500	2.83	707	4.01	616	5.04	872	7.1
12"					999	5.67	1413	8.02

Table 15, from ANSI A40-1993 Standard, *Safety Requirements for Plumbing*. Reprinted with permission.

1.93 feet per second (fps). However, this same 4″ drain line, if installed at ¼″ of slope per foot of pipe, will carry 53.4 gpm of fluid at a velocity of 2.73 fps.

For a construction manager, the slope or grade of a DWV pipe impacts the plumbing system in two ways.

First, most experts in the industry believe that a fluid velocity that is too high or too low will cause high maintenance expenses due to the constant clogging of the system. Certainly, this makes sense if the slope is small. The lower the slope, the slower the velocity of the fluid, the greater the chance that the pipe will not be properly scoured or cleaned during the operation of the system. Hence, a blockage develops that needs to be removed. This situation typically occurs approximately 30 minutes prior to your hosting a large dinner party introducing your new boss.

The problem with high velocities in a drainage system is a little harder to visualize. If the velocity is too high, the liquids in the drainage pipe may achieve a higher velocity than the solids, thus allowing the solids to be left behind. If the drainage line is not regularly used, these abandoned solids will cause a blockage in the drainage system.

As is the case with most aspects of the plumbing design, the slope or grade of a plumbing system is regulated by code. Most codes specify a minimum slope of ¼″ per foot of pipe for drainage pipes 3″ in diameter or smaller and a minimum slope of ⅛″ per foot of pipe for drainage pipes 4″ and larger. It has been my experience that drainage lines with a slope less than ⅛″ per foot of pipe usually have a long history of maintenance problems. Personally, I design plumbing systems such that the velocity within a drainage pipe does not fall below 2 fps, nor will the slope or grade exceed ½″ per foot of pipe. As always, it is imperative that you check your local plumbing code for the proper slopes and grades required within your jurisdiction.

The second impact that slope or grade has on the construction manager revolves around clearance and precedence. The fact that drainage pipes have slope means, of course, that the drainage pipes are constantly changing elevation. This important fact is usually driven home when the architect notices that the drain pipe serving the floor above is rather conspicuously installed below the lift-out ceiling . . . in the president's private office. Unfortunately, this situation is all too common in today's construction market. For example, if a particular drain line had a slope of ¼″ per foot of pipe and had a total run of 100′, it could easily be calculated that this drainage pipe requires 25″ of "fall." Because of economics, many modern buildings do not have 2′ of free space between the lift-out ceiling and the floor above. Hence, a conflict develops between the plumbing contractor and the general contractor.

As is usually the case, the arguments used by the various parties involved in this type of problem have several different perspectives. First of all, the plumbing contractor relies on the fact that most contracts contain language stating that systems that require gravity for proper operation have precedence over systems that don't require gravity. This statement, as it stands, is true. The second perspective comes from another contractor within the conflict who states that it is a lot less expensive for the plumbing contractor to offset their piping than it is to move a large air conditioning duct or similar obstruction. This is also true. Finally come the owners or their agent, the architect, who rely on contract language stating that the contractors are to coordinate their work to avoid these types of conflicts. This is a classic example of too many truths in one problem.

In this scenario, the alternatives are usually (1) lowering the ceiling, (2) raising the drain pipe by decreasing the slope, or (3) relocating the drain pipe to another location. All of these after-the-fact solutions are expensive, time consuming, and create hard feelings among all of the parties. The successful construction manager will learn to anticipate and avoid these pitfalls.

EVERY PLUMBING FIXTURE MUST HAVE A TRAP

The DWV system contains, besides the solid and liquid effluents from plumbing fixtures, sewer gases generated from these wastes. A plumbing trap is a device that uses a water seal that prohibits sewer gases from flowing out of the drain piping, through the plumbing fixture, and back into the occupied spaces of the building. This is accomplished without inhibiting the flow of effluent through the plumbing fixture.

Anyone who has encountered the rather distinctive odor of sewer gas knows how important that a trap is on a plumbing fixture. Sewer gas (a methane gas mixture) is a toxic, noxious, highly flammable gas. There are numerous examples of sewer gas accumulating in a building and igniting when some poor soul lit a cigarette or used a gas-fired appliance. The publicity accorded this type of event in the news media also serves to lower public perception of the plumbing industry. In fact, the whole construction industry suffers with this type of publicity.

The most common type of plumbing trap is referred to as a P trap, as the trap looks like the letter P lying on its back, as shown in Figure 2–1.

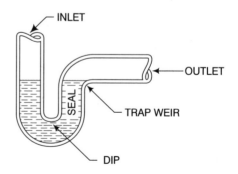

Figure 2–1 A typical trap section view.

Figure 1.2.59, *National Standard Plumbing Code*, 1993 Edition, p. 56. Reprinted with permission from the National Association of Plumbing Heating Cooling Contractors.

If you look under the kitchen sink or lavatory in your home, you will see a perfect example of this type of trap. The trap or seal is provided by a column of water, usually 2" to 4", located between the trap weir and the dip of the trap. This is what prevents the sewer gases from entering the building. Some types of plumbing fixtures, most notably water closets and urinals, have traps built into the fixtures. These are referred to as *integral traps*.

It should be stressed that there are a number of traps that are prohibited by one code or another. For example, most codes prohibit S traps, drum traps, and bell traps as these types of traps have a tendency, under the right conditions, to lose their seals. All codes, to the best of my knowledge, prohibit any trap that requires moving parts to maintain a seal.

The size of a fixture trap is also defined and regulated by code. Table 2–2, reprinted in part from the ANSI-A40 Standard, depicts common nonintegral trap sizes for most standard plumbing fixtures.

There are a number of special types of traps. One special type of trap common to commercial construction sites is the interceptor. An interceptor is a trap that separates out materials that are harmful to the plumbing system such as plaster, sand, mud, grease, oil, or other similar materials.

A solids interceptor separates the hazardous material from the normal wastes by producing an abrupt decrease in fluid velocity. This allows the unwanted solid materials, such as plaster, sand, or gravel, to settle out while letting the normal wastes continue to flow due to gravity. A grease trap is a special type of interceptor commonly required in commercial kitchens. In a grease interceptor, the grease is separated with a series of baffles or compartments to secure the grease in one location and prevent it from plugging or fouling the sanitary sewer. Typically, a separator is made specifically for a hazardous material. For example, a grease interceptor is a device different from a plaster interceptor.

A problem that is often associated with plumbing traps is evaporation. Over a short period, if the fixture is not used regularly, the water in the trap will evaporate, allowing sewer gases to enter the occupied spaces. As mentioned earlier, this is an invitation to disaster. This problem has particularly plagued floor drains, a special type of drain and trap that is installed in floors. The purpose of a floor drain is to allow cleaning of the floor or to receive the discharge from a condensate pan, relief valve, or other similar device. All homes have a floor drain next to the hot water heater or furnace, for example. Many architects, and even some engineers, believe that you can't have too many floor drains in a commercial building. The net result of this attitude is that many floor drains end up in areas where they receive very little use. Inevitably, the water in the traps evaporates, and the owner will immediately begin complaining about odors.

Many maintenance personnel believe a quick and easy solution to this problem is to pour a small amount of kerosene into the trap to keep the water from evaporating. Pouring a flammable liquid into a trap to keep a flammable

Table 2–2 Size of fixture traps for different types of plumbing fixtures.

WSFU[3,5]	Min. Size Cold[1]	Min. Size Hot[1]	Types of Plumbing Fixtures	Min. Trap Size[1]	Min. DFU Value	Min. Trap Arm or Fixture Drain[1]
2	½"	½"	Automatic clothes washer residential type with 2" standpipe	2"	3	2"
4	½"	½"	Commercial type	2"	3	2"
6	–	–	Bathroom group, flush tank	–	6	
8	–	–	Bathroom group, flushometer	–	8	–
2	½"	½"	Bathtub (with or without whirlpool or shower) private	1½"	2	1½"
4	½"	½"	Bathtub (with or without whirlpool or shower) public Continuous flow into drainage system 1 gpm = 2 DFU	1½"	4	1½"
1	½"	½"	Bidet	1¼"	1	1½"
2	½"	½"	Dental unit or cuspidor (with or without aspirator)	1¼"	1	1½"
2		¾"	Dishwasher (commercial restaurants)	1½"	2	2"
2		¾"	(With grease interceptor used as a trap)	2"	3	2"
1.5		½"	Dishwasher (residential type)	1½"	2	1½"
1	⅜"		Drinking fountain bubbler	1¼"	0.5	1¼"
–			Floor drain	2"	3	2"
–			Floor drain	3"	5	3"
–			Floor drain	4"	6	4"
	½"		Floor drain with trap primer (same as above)			
2	½"		Food waste grinder (commercial use)	2"	3	2"

[1]The sizes listed are the minimal inside diameter (i.d.) of the tube or pipe to each faucet or fixture.
[2]The fixture supply pipe shall be full-sized to a point within 30″ developed length of the fixture connection and within the same room as the fixture connection. The supply tubing or pipe used to make the final connection between the fixture supply pipe and the fixture may be reduced.
[3]Water supply fixture unit values listed are for total water demand for fixtures with both hot and cold water. The separate hot and cold supply fixture units values may be taken as three-fourths of the total water demand.

WSFU[3,5]	Min. Size Cold[1]	Min. Size Hot[1]	Types of Plumbing Fixtures	Min. Trap Size[1]	Min. DFU Value	Min. Trap Arm or Fixture Drain[1]
2	½"	½"	Laundry tray 1, 2, or 3 compartments	1½"	2	1½"
1	½"[2]	½"[2]	Lavatory (private use)	1¼"	1	1¼"
2	½"[2]	½"[2]	Lavatory (public use)	1¼"	1	1¼"
–	–	–	Mobile home park traps (one for each mobile home)	3"	6	3"
			Receptor (floor sinks) maximum flow rate from indirect waste pipes			
–	–	–	4.80 gpm	1½"	1.5	1½"
–	–	–	9.70 gpm	2"	2	2"
–	–	–	17.60 gpm	2½"	3	2½"
–	–	–	28.60 gpm	3"	4	3"
	–	–	57.00 gpm at ¼ grade	4"	6	4"
			Also see sinks with flush rims Sillcock, wall hydrant, or hose bib			
–	½"	–	½" with backflow preventer continuous flow 3 gpm[4]	–	–	–
–	¾"	–	¾" with backflow preventer continuous flow 4 gpm[4]	–	–	–
2	½"	½"	Shower compartment (private use)	2"	2	2"
2	½"	½"	Shower (gang) per head	2"	1	2"
2	½"[2]	½"[2]	Sink (residential, with or without food waste grinder and/or dishwasher) one trap	1½"	2	1½"
1	½"[2]	½"[2]	Sink (bar) private	1¼"	1	1¼"
			Sinks (surgeon, dentist)			
2	½"[2]	½"[2]	1 compartment	1¼"	1.5	1½"
2	½"[2]	½"[2]	2 compartments	1½"	2	1½"
2	½"[2]	½"[2]	3 compartments, faucet	1½"	3	2"
3	½"[2]	½"[2]	3 compartments, 2 faucets	1½"	3	2"

[4]Continuous flow gallons per minute (gpm) to lawn sprinklers, air conditioning, industrial uses, etc., shall be added to the gpm for fixture in that part of the water supply system where the connection occurs and in the total water demand.

[5]For fixtures not listed in the table, water supply fixture units (WSFU) values shall be determined (1) by comparison with a fixture consuming water in similar quantities and at similar rates or (2) be calculated at their maximum demand, but in no case less than.

Table 2-2, *continued*

WSFU[3,5]	Min. Size Cold[1]	Min. Size Hot[1]	Types of Plumbing Fixtures	Min. Trap Size[1]	Min. DFU Value	Min. Trap Arm or Fixture Drain[1]
			Sink (commercial, kitchen, scullery, with or without grinder)			
2	½"	½"	1 compartment	1½"	3	2"
2	½"	½"	1 compartment with grease interceptor as trap	2"	3	2"
2	½"	½"	2 compartments	1½"	3	2"
2	½"	½"	2 compartments with grease interceptor as trap	2"	3	2"
–	–	–	3 compartments	1½"	4	2"
–	–	–	3 compartments with grease interceptor as trap	2"	4	2"
2	½"	½"	3 compartments with 1 faucet	–	–	–
4	½"	½"	3 compartments with 2 faucets	–	–	–
2	½"	½"	Sink (cast room, plaster, etc. with interceptor)	2"	3	2"
3	½"	½"	Sink (service 2") P-trap	2"	3	2"
3	½"	½"	Sink (service 3") P-trap	2"	3	2"
10	1"	–	Sink (clinic with flush rim)	Integral	6	3"
4	½"	½"	(Faucet above clinic sink)	–	–	–
2	½"	½"	Sink combination sink and tray (each trap)	1½"	2	1½"
2	½"	½"	Sink, wash, multiple faucets (each faucet, water only)	1½"	2	1½"
3	¾"	¾"	Foot-operated with spray pipe	1½"	2	1½"
2	½"	½"	Sitz bath	1½"	2	1½"
2	¾"	–	Wall urinal washout ¾ spud/flushometer	Integral	4	2"
2	½"	–	Wall urinal washout ¾ spud/automatic siphon tank	Integral	4	2"
2	½"	–	Wall urinal siphon jet ¾ spud with automatic siphon tank	Integral	4	2"

[6]For the purposes of the table, a bathroom group consists of not more than one water closet, one lavatory, one shower stall, or not more than one water closet, two lavatories, one bathtub, and one separate shower stall.
[7]The sizes listed are the nominal inside diameter of the tube or pipe to each faucet.
[8]Drainage piping serving batteries of appliances capable of producing continuous flows shall be sized to provide for peak loads. Clothes washers in groups of three or more shall be rated as six drainage fixture units each for the purpose of common waste pipe sizing.

Table 2-2, continued

WSFU[3,5]	Min. Size Cold[1]	Min. Size Hot[1]	Types of Plumbing Fixtures	Min. Trap Size[1]	Min. DFU Value	Min. Trap Arm or Fixture Drain[1]
10	1"	–	Wall urinal blowout 1¼" and 1½" spud/flushometer	Integral	6	2 or 3"
5	¾"	–	Wall urinal siphon jet ¾ spud/flushometer	Integral	4	2"
10	1"	–	Wall urinal siphon jet 1¼" spud/flushometer	Integral	6	3"
10	1"	–	Urinal, pedestal	Integral	6	3"
3 6	½" ½"	– –	Water closet, one piece, minimum flow pressure as required per manufacturer Private, round front or elongated Public, elongated only	Integral Integral	4^{10} 6	3" 3"
3	⅜"	–	Water closet, close coupled or high tank private, round front	Integral	4^{10}	3"
3	⅜"	–	Water closet, close coupled or high tank, private, elongated front	Integral	4^{10}	3"
5 10	1" 1"	– –	Water closet, siphon jet with flushometer valve, minimum flow pressure, 15 PSI Private, round front or elongated Public, elongated only	Integral Integral	4^{10} 6	3" 3" 3"
3	⅜"		Water closet, flushometer tank, 20 PSI	Integral	4	3"
5 10	1" 1"	– –	Water closet, blowout, minimum flow pressure, 25 PSI Private Public	Integral Integral	4^{10} 6	3" 3" 3"

Legend:
WSFU = Water Supply Fixture Units
Min. = Minimum
DFU = Drainage Fixture Units
Superscripts (e.g., WSFU[3,5]) refer to Footnotes.

[9]For fixtures not used in this table, drainage fixture unit (DFU) valves shall be determined (1) by comparing the fixture to a fixture discharging waste in similar quantities and at similar rates or (2) by using the DFU comparison table below:
[10]Tank-type water closets shall be computed as six drainage fixture units when determining septic tank sizes.
[11]Trap size shall not be increased where the fixture discharge may be inadequate to maintain self-scouring properties.
[12]For a continuous or a semi-continuous flow into a drainage system, such as from a pump (sump or ejector), air-conditioning equipment, or a similar device, two fixture units shall be allowed for each gpm of flow.
Table 2, ANSI A40-1993 Standard, *Safety Requirements for Plumbing.* Reprinted with permission.

Table 2–2, *continued*

Fixture Sizes, in.	Number of Fixture Units	
	Private Use	Public Use
3/8"	1	2
1/2"	2	4
3/4"	3	6
1"	6	10

Fixture Outlet or Trap Size	DFU Value
1¼"	1
1½"	2
2"	3
2½"	4
3"	5
4"	6

gas from seeping out is not a good idea. I can foresee a "closet" smoker, not wanting to be caught in the act of smoking, quickly throwing a cigarette butt into an open floor drain. The good news is that the smoker would probably decide to quit smoking—once the fire is extinguished.

If it is absolutely necessary to install a floor drain in any area of low usage, I would recommend one of two possible solutions. The first would be to use a deep seal P trap. This type of trap contains a larger reservoir of water than a standard trap, thus it does not evaporate as quickly as other traps. The second would be to install a trap primer. This is a device that automatically adds water to a trap as necessary to maintain a minimum trap seal. If all else fails, I have found that adding a little vegetable oil to the trap will at least slow down the evaporation rate.

EVERY TRAP MUST HAVE A VENT

This is the last of the major fundamentals of plumbing and, in many respects, it is one of the most important.

To fully understand the concept of venting, it is first necessary to understand that a properly sized drain pipe contains wastes (both liquid and solid) and air. For a horizontal drainage pipe, an optimal design calls for the pipe to flow half full of waste and half full of air. In an optimal design for a vertical drain line, the pipe will be approximately one-third full of waste with the remaining space occupied by air. It should be observed, that in the case of a vertical drain pipe, the liquid flow is on the wetted perimeter of the pipe, making the flow look similar to a doughnut. All of the tables used in this text to size drain, waste, and vent piping are based upon this concept of an optimal design. These flow conditions are depicted in Figure 2–2.

The fact that the drainage system contains both wastes and air is significant for two different reasons. First, a properly designed plumbing system "breathes" air into and out of the system, in response to drainage flows in the system. It follows, then, that we must provide some means to allow air to enter and leave the drainage system. Second, it must be remembered that the wastes in a drainage system flow due to gravity, not a pressure differential. Therefore, it is important that a neutral pressure be maintained within the drainage piping so that the flow of wastes is not obstructed due to a pocket of

Figure 2–2 Flow of waste in horizontal and vertical drain pipe.

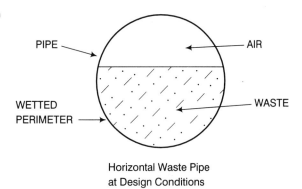

Horizontal Waste Pipe at Design Conditions

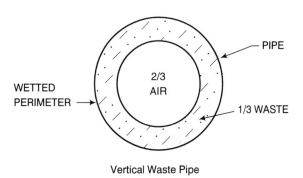

Vertical Waste Pipe at Design Conditions

high or low pressure. For this reason, most plumbing systems are referred to as *nonpressurized* systems.

The venting system is designed to circulate air throughout the plumbing system in order to achieve three main goals:

1. Remove dangerous odors and gases from the plumbing system
2. Prevent positive pressures from being developed within the plumbing system
3. Prevent negative pressures from being developed within the plumbing system

Venting the noxious odors and flammable gases out of the plumbing system is a clear goal of any venting system. This goal is achieved by ensuring that the vent system terminates outside the building, to the atmosphere, usually above the roof line. Most codes require this vent termination to be at least 12" above the roof, principally so the vent is not clogged or closed with snow, water, or debris. The plumbing vent should also not terminate within 10' horizontally of any fresh air intake, window, door, or other such intake into the building, unless the vent is extended at least 3' above the intake. It only makes sense, after you have carefully made provisions to remove the noxious gases, not to allow these same gases easy access back into the building.

Also, in cold weather climates, the diameter of the vent termination should not be less than 3" in diameter, so that the vent is not blocked by frost in the winter months. This is a very real threat, inasmuch as the air discharging from a vent stack is usually quite warm and contains a lot of moisture.

The problem of plumbing systems developing either too much pressure or even developing a negative pressure in the piping is a little more difficult to visualize. The diagram in Figure 2–3 may help.

When design flow is exceeded (and this will happen, even in a properly designed system), the flow of liquid in a vertical pipe becomes a completely filled cylinder, as opposed to a thin wall cylinder, as shown in Figure 2–3. This "slug" of water, traveling down a vertical pipe, will create a positive pressure zone in front of the cylinder and a negative pressure zone behind the cylinder. As it comes to a horizontal branch, the branch piping, and all of the fixtures attached to the branch, are subjected to the extra pressure.

Although it is easy to see this phenomenon, the initial reaction of most people is that this is not a "big deal." This is exactly how most plumbing problems start.

From our discussion on plumbing traps, remember that most traps contain a water seal of approximately 2" of water. From physics, recall that a column of water 12" high, at 68°F, exerts 0.4335 lbs per sq. in. (psi) of pressure at its base. Therefore, a plumbing trap with a 2" water seal will provide $2/12$ (2" of water seal / 12" of water per foot) × 0.4335 psi, or approximately 0.072 psi of pressure protection. Clearly, it does not take a lot of pressure to compro-

Figure 2–3 Positive and negative pressures in a stack.

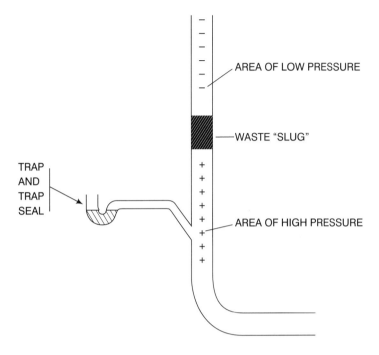

mise the integrity of the plumbing trap. Now it should be easy to see how a simple slug of water traveling down a vertical pipe can generate plenty of excess pressure. Enough excess pressure, in fact, to literally blow the water seal out of the trap and onto anyone unfortunate enough to be standing in front of the fixture. This phenomena is called, appropriately enough, *trap blowout*. Virtually every plumbing contractor has at least one very humorous story about a bank president receiving a "free shower" while standing in the bathroom of a high-rise bank building that is not properly vented.

The exact opposite can occur as the slug of water passes the connection to the branch. The negative pressure on the trailing edge of the slug can literally draw the water out of the trap seal and into the branch drain line, where it drains away like any other fluid. When this occurs, sewer gases have an open path through the fixture and into the occupied space. This phenomenon is called *trap siphonage*.

A venting system is designed to prevent these kinds of problems. Because the plumbing system contains traps that usually contain a 2" water seal, most venting tables, including Table 24 of the ANSI-A40 Code (see Chapter 6, Table 6–4), are based on a friction loss of 1" of water column (W.C.). Therefore, when a venting system is designed, the air within the drainage system is forced to "look" at two different paths: the path through the fixture trap, which has a resistance of 2" of water column; and the path through the venting system, which has a resistance of 1" of water column. Air, like water, always follows the path of least resistance; hence, any excessive pressure differential will travel through the vent pipe to the atmosphere, thus always protecting the trap.

This explains why we must increase the size of the vent pipe to compensate for greater distances within the system. Look at it this way: The more pipe and fittings along the path in the vent system, the more friction loss (i.e., resistance) the air will "see" in selecting a path to travel. Increasing the diameter of the pipe decreases the amount of resistance that the air will see. Therefore, the longer the vent system, the greater the diameter of the vent pipe. This is the theory behind all vent piping charts and tables.

This also explains why most codes regulate the distance between the trap and the trap's vent connection. This distance is usually referred to as the *trap arm*. If the vent is to protect the trap adequately from pressure differentials, it is necessary to ensure that the vent is close enough to the trap so that any pressure differential that exists can see two paths. For the ANSI-A40 Standard, this regulation on trap arm length is found in Table 2–3.

All of the major plumbing codes also require that the vent piping be installed with a certain grade or slope. On the surface, this seems unnecessary. After all, the vent piping contains only air. Although this statement is true, remember that the air in the vent piping is warm and contains a lot of moisture. As this vent pipe passes through unconditioned spaces, the wall of the pipe will cool, which, in turn, causes the moisture to precipitate out of the air. As a result, you often have a small amount of water continuously flowing in the bottom of the vent piping system. This explains why the vent piping should be gradually sloped back toward the drain or stack to which it connects. Usually, the grade is no more than ⅛" to ¼" per foot of piping. It also means that you need to avoid inadvertently building traps into the vent piping. Clearly, if you have a series of fittings that produce a trap, the water in the vent piping will eventually fill the trap with water. Once the trap is full, air flow throughout that section of venting piping is prohibited, and the fixture traps are suddenly unprotected from pressure surges.

Table 2–3 Maximum length of trap arm.

Size of Fixture Drain, in.	Distance–Trap to Vent, ft
1¼	2.5
1½	3.5
2	5
3	6
4	10

Table 19, ANSI A40-1993 Standard, *Safety Requirements for Plumbing*. Reprinted with permission.

Drain, Waste, and Vent Fundamentals

One last regulation to consider in the installation of the vent piping system is to ensure that the vent pipe from a plumbing fixture always rises above the flood level rim of that particular fixture. In other words, the vent piping should rise up vertically at least 6″ above the flood level rim before it turns horizontally to connect to another vent pipe. This will keep the vent pipe from acting like a waste or drain pipe in the event that the drain line from the fixture becomes plugged. If the drain line does become plugged, you want the excess waste to spill out of the plumbing fixture and not invisibly back up the vent pipe. Excess waste spilling out of a plumbing fixture always tells you (right away, I might add) that you have a serious problem that needs to be solved. If the excess waste simply backs up into the vent system, you would never know you had a serious problem until you had a major catastrophe with the system as a whole.

A few words about specialty vents is also in order. Undoubtedly, you will uncover instances of "wet" venting and combination waste/vent systems in many, if not most, buildings. The riser diagram in Figure 2–4 shows a simple wet venting system that is very common.

Figure 2–4 Riser diagram.

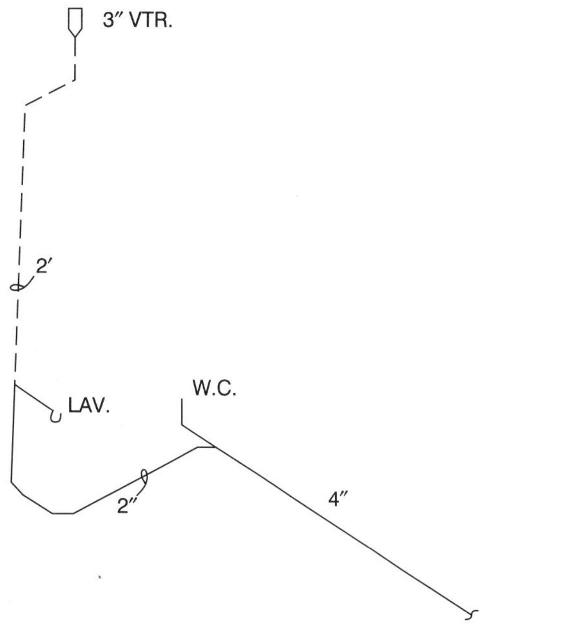

No Scale

In this illustration, the lavatory drains into the vent line for the water closet. Therefore, this vertical line is both a drain line for the lavatory and a vent line for the water closet. This, in turn, means that the water closet vent line periodically will have liquid flowing in it, which is why it is called a "wet" vent. This is also why the vertical drain serving the lavatory is 2" in size (the vent size required for a water closet) and not 1½" (the drain size required for a lavatory by Table 2–2). This works very well for single bathrooms, as it would be very rare to use the lavatory and the water closet at the same time. Thus, both fixtures are adequately drained and vented. Most codes allow wet venting of certain fixtures in certain applications.

Most codes also allow what is called *combination waste and vent systems*. This is where the waste or drain line is oversized to the extent that one pipe has enough volume to carry an adequate amount of waste on the bottom of the pipe and an adequate amount of air for venting purposes on the top of the pipe. Even though these systems are commonly used, they should be used with some care to ensure that proper venting is always being provided. Because a combination waste/vent system is often less expensive to install, good judgment can become clouded by the temptation of a short-term economic gain. As an example of a good application of combination waste/vent systems, most codes allow a type of combination waste/vent system for branch piping that serves only floor drains in commercial buildings.

It is also possible, and even desirable, to have a vent serve more than one trap. These types of vents, called *loop or circuit vents* (see Figures 2–5 and 2–6), are generally limited to a certain number of floor-mounted plumbing fixtures such as water closets, floor drains, and showers. These types of vents work very well, particularly for banks or rows of similar fixtures. The venting of these fixtures is also based upon diversity—the concept that not all fixtures in a row will be used at exactly the same time.

One last type of vent to be aware of is what is generally referred to as a *"quickie" vent* (see Figure 2–7). There are some plumbing installations, such

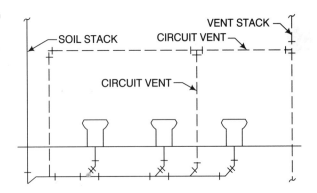

Figure 2–5 An example of a circuit vent

Figure 1.2.65, *National Standard Plumbing Code*, 1993 Edition, p. 60. Reprinted with permission from the National Association of Plumbing Heating Cooling Contractors.

Figure 2–6 An example of a loop vent.

Figure 1.2.69, *National Standard Plumbing Code,* 1993 Edition, p. 61. Reprinted with permission from the National Association of Plumbing Heating Cooling Contractors.

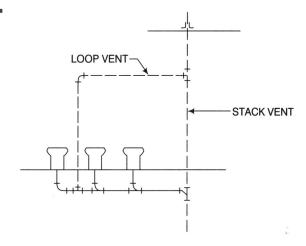

as a sink installed in an island counter, where it is virtually impossible to install a proper vent. Therefore, most codes allow, with some restrictions, automatic vent devices or quickie vents. These devices typically use a diaphragm or a spring to allow air to enter the venting system in the presence of a negative pressure condition near the fixture trap. Quickie vents are certainly better than no vent; however, they should be used with some caution. They do have moving parts, and, as we all know, anything with moving parts eventually fails. Therefore, quickie vents should be installed only where they are accessible for repair and replacement. Additionally, they should be regularly inspected to ensure that they are working properly. Finally, note that quickie vents do not allow for air to leave the venting system; therefore, they do not protect against overpressurization.

Figure 2–7 Cutaway view of quickie vent.
Courtesy of J & B Products.

REVIEW QUESTIONS

1. What are the three main purposes of a plumbing vent system?
2. Why is it important to ensure that the slope, or grade, of a drain or waste line is neither too high nor too low?
3. What is the velocity of the wastewater in a 3" pipe installed with a slope of ⅛" per foot? Is this acceptable? Why or why not?
4. What is the purpose of a plumbing trap?
5. What is the purpose of an interceptor? How does it work?
6. Why must the vent line serving a trap rise 6" above the flood level rim of the fixture?
7. What is the difference between a circuit vent, a loop vent, and a wet vent?
8. Why does a vent line increase in size as the total developed length (TDL) of the vent line increases?
9. Define a trap arm. What is the minimum trap arm for a fixture drain of 1½"?
10. What is the minimum trap size for a service sink?

CHAPTER 3

Materials

SELECTION CRITERIA

There are a number of different materials that are commonly used in the plumbing industry. As a construction manager, it is important to be aware of the different materials available and the advantages and disadvantages of each type of material.

The following criteria are used by most design engineers in selecting a material for the plumbing system:

   1. **Corrosive nature of the fluid being conveyed** Some wastes are very corrosive, such as the acid wastes found in many laboratories. The waste and vent piping for these types of systems must be made of special materials that resist corrosion.

   2. **Temperature and pressure of fluid being conveyed** As we have already discussed, most waste and vent systems are gravity flow systems; that is, they are nonpressurized systems. However, there are a few situations that may develop, such as the discharge piping from a sewage ejector, where the waste piping can be pressurized. In these cases, special consideration must be given to the pressure rating of the piping. Additionally, the wastes from some equipment, particularly boilers, may be at a high temperature. Even though most codes prohibit any waste over 180° from entering the plumbing system, 180° water can seriously damage some types of piping materials.

3. **Piping materials that are allowed by code** Some local plumbing codes are very restrictive and others are very lax on what types of materials are allowed. At the risk of sounding like a broken record, there is simply no substitute for studying the local code in your jurisdiction.

4. **Availability of piping material** Some piping material is not available in certain sizes. For example, cast-iron soil pipe is not available in 1¼″ diameter. If the plans call for this size of piping, you either have to provide a different material, such as DWV grade copper, or you have to increase the piping size to 1½″ in diameter.

5. **Cost of piping material** Needless to say, cost is an important criterion to keep in mind when selecting a plumbing material. Considerable savings can be achieved by switching from an expensive piping material, such as copper, to a less expensive material, such as plastic.

6. **Cost of installation of piping material** This goes hand in hand with the discussion on material costs. For example, a glued joint, as used in most plastic systems, will be quicker and less expensive than the soldered joint used in most copper systems. The number and type of piping supports required by the code is also an important economic consideration.

7. **Miscellaneous criteria, such as noise considerations** This is one of those famous "catch-all" phrases where we include items that are not always fully understood. There are, however, many other special considerations to remember when selecting a piping material. For example, most building codes prohibit any combustible materials from being placed within a return air plenum, such as is often found above a lay-in ceiling. If your project has such a return air plenum, make sure that your plumbing contractor is not planning on using a combustible material, such as PVC plastic, for the drain, waste, and vent piping in this plenum. Another consideration is noise. Some materials, such as some of the thinner wall plastics, do not do a good job of absorbing the sounds generated within a plumbing system. I was the mechanical project manager on a high-rise condominium project that had a 6″ PVC sewer riser which ran vertically through a number of the very expensive luxury condominiums. As soon as the project opened, complaints started rolling in about the "Niagara Falls" effect echoing through their unit. These are the types of conversations that you want to avoid.

MATERIAL TYPES

Plumbing materials can be broken down into the following classes:

1. Plastics
2. Cast-irons and carbon steels

3. Nonferrous materials
4. Stone-based materials

Plastics

Plastics are one of the most popular of all plumbing materials. Although plastics have been widely used for drain, waste, and vent piping, they are starting to gain acceptance for domestic hot and cold water piping as well. Many local natural gas utilities also use a type of plastic pipe for the underground gas main piping. Additionally, fire sprinkler contractors are beginning to use plastics for the sprinkler piping in residential and commercial projects.

Plastic pipe and fittings are usually described by the pressure rating of the pipe. For example, Schedule 40 PVC pipe (generally considered "standard" weight) defines a set of pipe diameters that all have approximately the same pressure rating. To achieve this, the pipe wall thickness increases as the pipe diameter increases. For plastic pipe used in underground storm and sanitary drainage, the pipe is often referred to by its SDR rating. SDR stands for "standard dimensional ratio" which is the specific ratio of the average specified outside diameter to the minimum wall thickness. This applies to outside, controlled diameter plastic pipe. Under this system, Schedule 80 pipe is sometimes referred to in the industry as "extra heavy" weight, as the wall thickness is significantly larger than the standard wall thickness of Schedule 40.

Most plastic systems use a glued type of joint; however, mechanical couplings, fused joints, or even threaded joints (on Schedule 80 pipe) are also common. Most glued joints depend on the application of a primer, then the application of the adhesive. Because it is almost impossible to tell the primer from the adhesive, many codes require the primer to be a bold color, such as purple. In this manner, the inspector can easily see if the primer was properly applied. In addition to a visual test, most codes also require a hydrostatic test that imposes a water pressure of 10′ of head (height) on the system under test (see Figure 3–1).

Generally speaking, plastics are usually quite inexpensive to buy and install. They typically have good to excellent corrosion properties and are available in a wide range of sizes in most communities in the United States. Plastics also have a fairly decent thermal resistance. This means that in some applications, such as condensate drains off of cooling coils, plastic pipe can be used without being insulated to prohibit condensation from forming on the outer wall of the pipe. This would not be the case, for example, with copper piping, which has poor thermal resistance.

On the downside, most plastics do not respond well to heat; therefore, you should make every effort to see that the effluent temperature discharging into a plastic system does not exceed 140°. As previously mentioned, many plastics are considered a combustible material, which precludes their use in

Figure 3-1 (a) Typical plastic solvent cement joint for DWV or water distribution. (b) Typical plastic elastomeric gasket joint for DWV. (c) One type of stainless steel clamp and elastomeric gasket joint used for DWV plastic piping—shielded coupling shown.

Figures 4.2.14a, 4.2.14b, 4.2.14c, *National Standard Plumbing Code,* 1993 Edition. Reprinted with permission from the National Association of Plumbing Heating Cooling Contractors.

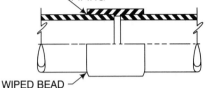

(a)

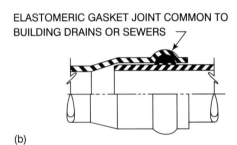

(b)

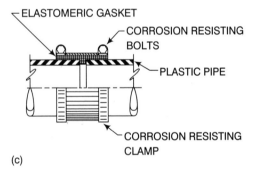

(c)

return air plenums and other similar fire-restricted areas. Also, noise problems with plastics can sometimes create a problem. Finally, plastics generally have a high thermal expansion value, which means they tend to expand or contract as heat is applied or removed from the plastic. Accordingly, many codes limit plastic piping to buildings of three stories or less. Personally, I believe that plastic DWV systems can be used successfully in taller buildings, if the problems of expansion and contraction are properly addressed.

The major classifications of plastic piping and their principal functions are as follows:

1. **Acrylonitrile Butadiene Styrene (ABS)** Generally used in above-ground DWV systems.
2. **Chlorinated Polyvinyl Chloride (CPVC)** Generally used in domestic hot and cold water systems.
3. **Fiberglass** Usually used for corrosive wastes or fuel oil piping. Also used as a secondary containment pipe for acid waste systems.
4. **Polybutylene (PB)** Generally used for domestic hot and cold water piping. This material is also used for fire sprinkler piping. It needs to be noted that polybutylene piping, because of some recent failures that were highly publicized, is currently being scrutinized by the industry. As a result, there are a number of codes that strictly prohibit polybutylene pipe and fittings. However, not all of the facts are known concerning the failures of polybutylene, so keep an open mind regarding its future use in the industry.
5. **Polyethylene (PE)** Widely used for underground piping applications for natural gas and domestic water. Also, cross-linked polyethylene (PEX) and cross-linked polyethylene/aluminum/cross-linked polyethylene (PEX-Al-PEX) are new developments in domestic water piping that are gaining acceptance in some jurisdictions.
6. **Polypropylene (PP)** Often used for acid waste piping, process piping, and high-purity water systems.
7. **Polyvinyl Chloride (PVC)** The "Big Dog" of plastic plumbing systems. It is widely used in drain, waste, and vent systems; high-purity water systems; corrosive waste systems; and process piping.
8. **Polyvinylidene Fluoride (PVDF)** Typically used in high-purity water systems and for some types of process piping.

Cast Irons and Carbon Steels

This is another very popular category of plumbing material, dominated by the cast-iron products.

Cast-iron pipe comes in two different grades: standard weight (SV) and extra heavy (XH). Most codes allow standard weight cast-iron pipe and fittings to be used in most drain, waste, and venting applications both above ground and below ground. Extra heavy cast-iron pipe and fittings are rarely specified and used in construction today; hence, their availability is limited. There is a special type of cast-iron product with a high silicon content that is sometimes specified for acid waste applications and is referred to by its trade name "Duriron." Joints for cast-iron piping are usually hub-and-spigot for underground applications. With this type of joint, a rubber neoprene gasket is used to make the watertight joint. Historically, in lieu of the neoprene gasket, this same joint was made with an oakum packing and a poured lead seal on top. You will find this type of joint in many of our older buildings. For above-ground applications, a no-hub joint is commonly used. This provides a watertight joint through the use of a neoprene coupling that is "banded" to each side of the joint. As for the plastic systems, most codes require a hydrostatic test of 10′ of head on each section of piping. Remember that this test of 10′ of head represents only 4.3 psi of pressure. The vast majority of these joints, like the systems themselves, are considered nonpressurized (see Figure 3–2).

The carbon steels, particularly galvanized carbon steel pipe, have been used for many years in a variety of drain, waste, and vent applications. Today, steel pipe is usually used for fire sprinkler piping and natural gas piping. Because of cost and corrosion problems, carbon steel is rarely used in DWV systems, although it is still permissible in most codes. The carbon steels come in a wide

Figure 3–2 (a) One example of a hubless pipe joint—shielded coupling shown. (b) Typical hubbed joint with a compression gasket.
Figure 4.2.11.2a and 4.2.11.2b, *National Standard Plumbing Code,* 1993 Edition. Reprinted with permission from the National Association of Plumbing Heating Cooling Contractors.

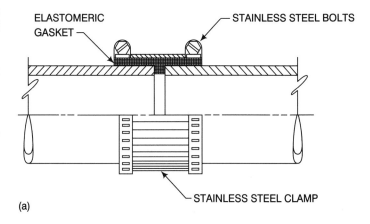

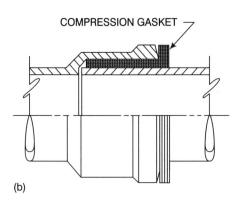

variety of thicknesses, from Schedule 5 (the thinnest wall) to Schedule 160 (the heaviest wall). Schedule 40 is generally considered standard, and Schedule 80 is generally considered extra heavy, depending on the diameter of the pipe. Carbon steel systems are typically joined with either threaded joints or, for piping 2½″ and larger, with a welded joint which utilizes flanges (see Figure 3–3. Most fire sprinkler systems, however, use what is called a *rolled or grooved joint*, which uses a groove cut or rolled into the pipe end and a special coupling to connect the two ends. Because of the popularity of the manufacturer, these are sometimes improperly referred to in the industry as *Victaulic joints*.

Usually kept as a separate class are the stainless steel materials. Obviously, they are usually used for very corrosive wastes or for high-purity applications. Like the carbon steels, they also come in a wide variety of grades or schedules. Stainless steel tubing or piping can be joined with threaded joints, welding, or a type of mechanical joint or clamp.

The final material in the cast-iron and carbon steel category is ductile iron pipe and fittings. Ductile iron is commonly used for underground domestic water main piping 3″ and larger. When used for this application, the ductile iron pipe will have a thin coating of cement inside the piping, and either a bitumastic tar exterior coating or perhaps a PVC exterior coating. Ductile iron pipe and fittings generally use what is called a "push-on" type of joint which utilizes a neoprene gasket inside a hub on one end of the joint. Flanged and bolted joints (sometimes referred to as a *mechanical joint)* are also used in joining ductile iron pipe as is the rolled and grooved coupling discussed before (see Figure 3–4).

As a class of materials, cast-iron and carbon steel are still very popular, although their popularity is slipping. The cast-iron and steels handle both pressure and temperature much better than the plastics. If you are going to be producing a high temperature waste, for example, you may well want to select one of these materials. The cast-iron and steels are also nonflammable. Therefore, they are suitable for use in a return air plenum or for the penetration of a fire wall, whereas most plastics would not be suitable. The cast-iron

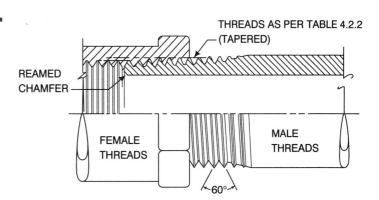

Figure 3–3 Typical threaded joint.
Figure 4.2.2, *National Standard Plumbing Code*, 1993 Edition. Reprinted with permission from the National Association of Plumbing Heating Cooling Contractors.

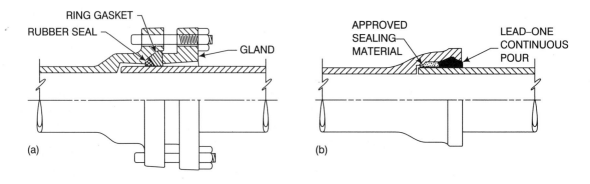

Figure 3-4 (a) One example of a mechanical joint for cast-iron water pipe. (b) Typical cast-iron water pipe—hubbed (caulked).
Figures 4.2.11.3 and 4.2.1.2, *National Standard Plumbing Code,* 1993 Edition. Reprinted with permission from the National Association of Plumbing Heating Cooling Contractors.

and steels also offer superior strength and, thus, are very popular in underground applications, particularly under roads and driveways. They also handle abuse much better than most other materials.

The downside usually revolves around cost. The superior strength and weight also mean that they are more expensive to install than their main competitor, plastic systems. Many of the carbon steels also have a problem with corrosion with respect to certain chemicals and fumes.

Nonferrous Metals

This fancy title basically refers to copper and brass piping, although lead and aluminum piping are generally included as they are also nonferrous metals.

Lead piping was long used for the underground domestic water main piping from the city main to the meter inside a home. In fact, many older homes still have and use the original lead water service pipe. Our new understanding of the danger of lead in the drinking water has, for the most part, relegated lead piping to a very minor role as a specialty material for corrosive drain and waste piping.

Aluminum piping is rarely, if ever, used in plumbing systems as such. It is most often found in specialty piping applications, such as cryogenic piping (liquid nitrogen, for example).

Brass piping has long been used to convey high temperature and high pressure water in domestic water systems. It is also used extensively as a trim material. For example, chrome-plated brass piping is commonly specified for exposed water piping in commercial buildings. As most alloys of brass also contain lead, the use of brass piping has, predictably, been falling.

By far, one of the most popular materials for use in plumbing systems is copper pipe and fittings. It is used for domestic water piping; drain, waste, and vent piping; and medical gas piping.

Copper piping comes in four different grades or wall thicknesses. Further it can be purchased in either "hard" lengths or in annealed or "soft" rolls of tubing. Most piping is identified by stenciling on the outside of the pipe; however, the stenciling itself is also color coded. Type M pipe has red stenciling; Type L has blue stenciling; Type K has green stenciling; and DWV has yellow stenciling. Types M, L, and K are used for domestic water piping and in pressure situations. Type DWV pipe is used only on drain, waste, and vent piping and is considered a nonpressurized pipe. Copper pipe is joined by either soldered joints, brazed joints, or a type of mechanical joint, such as a slip joint or a flared type of joint. Soldered joints on domestic water piping must be made with lead-free solder, which is usually comprised of a tin-antimony mixture. Soldered joints, made at temperatures below 900°, must use a flux compound on the joint to enable the solder to penetrate the joint completely. Brazed joints, made at temperatures above 900°, do not use a flux to distribute the filler. Many engineers prefer the brazed joint, believing that a defective brazed joint will manifest itself right away, whereas a poor soldered joint may not leak for a significant amount of time because the flux acts as a temporary plug for a void or hole in the soldered joint.

Slip joints and compression joints are similar in that both use a threaded nut in compression to form a joint. The slip joint uses a washer and a locknut to achieve a tight joint whereas the flared fitting uses a flare nut which compresses the flared portion of the annealed ("soft") copper tubing (see Figures 3–5 and 3–6).

It is common industry practice to specify a "soft" or annealed copper for installation underground. A soft copper pipe will handle the stresses and forces encountered in underground applications better than the hard copper pipes which may be subject to cracking or "kinking."

Copper piping is used extensively in plumbing systems primarily due to the ease of installation, which invariably translates into lower total system costs. Copper piping is quite durable and will take quite a bit of abuse. As with the carbon steels, copper piping is not flammable, thus it is suitable for use in return air plenums and fire wall penetrations.

Copper pipe and fittings, although easy to install, are more expensive than comparable plastic materials. Another problem with copper piping concerns electrolysis, the transfer of electrons between dissimilar metals. With copper pipe, you have to be aware of connections to the types of piping, particularly carbon steel piping. The transfer of electrons from one pipe to another pipe will cause corrosion and pitting in the donor pipe (usually one of the carbon steels).

Figure 3–5 (a) Typical soldered joint. (b) Typical brass ring joint—it is not a slip joint if rigid when tightened. (c) Typical washer joint—slip joint.

Figures 4.2.4, 4.2.15a, and 4.2.15b, *National Standard Plumbing Code,* 1993 Edition. Reprinted with permission from the National Association of Plumbing Heating Cooling Contractors.

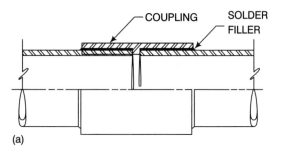

(a)

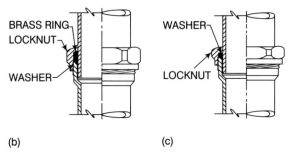

(b)   (c)

NOTE: WHEN A GROUND JOINT BRASS CONNECTION IS USED AND THE JOINT, WHEN ASSEMBLED, DOES NOT PERMIT FREE MOVEMENT, THEN THE JOINT NEED NOT BE ACCESSIBLE.

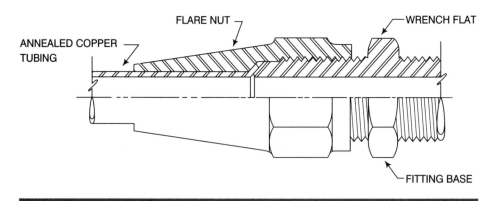

Figure 3–6 Typical flared joint.

Figure 4.2.5, *National Standard Plumbing Code,* 1993 Edition. Reprinted with permission from the National Association of Plumbing Heating Cooling Contractors.

Stone-based Piping

Stone-based piping covers a wide range of materials, from concrete pipe to fiberglass pipe. Most pipe of this type is manufactured in a mold and then is fired or cured to produce the final product.

Concrete pipe or reinforced concrete pipe (RCP) is widely used on underground sanitary and storm drainage systems outside the building walls, particularly for very large sizes. For example, RCP pipe can be purchased in diameters up to 144″. Concrete pipe uses a hub-and-spigot type of joint similar to cast-iron soil pipes. The joints on a concrete piping system can be made with either an O-ring gasket or with the use of mortar, where allowed by code (see Figure 3–7).

Clay tile pipe is also a very popular material for underground sanitary and storm mains outside the building wall. It has better corrosion resistance than concrete and is available in both standard weight and extra heavy weight. Joints are made with either a hub-and-spigot joint which uses a concrete grout or a push-on joint which uses a neoprene gasket. Unfortunately, clay pipe is far more susceptible to cracking and leaking over time than some of the other materials commonly used in underground applications. This disadvantage, coupled with the higher material and installation costs, has contributed to its decline in popularity. As a result, the use of clay tile in plumbing systems has been falling rapidly as plastic piping systems have become popular.

Asbestos-cement piping is also used for underground sanitary and storm drainage; however, it can also be used for underground domestic water piping if it has an epoxy lining to protect the water supply. Asbestos-cement piping uses a push-on joint which utilizes a neoprene gasket. It should be noted that asbestos-cement piping is not commonly used in the plumbing industry due to cost and the difficulty in installation. Additionally, all asbestos-bearing materials have been heavily regulated since the discovery of asbestos as a respirable health hazard in the 1970's. As a result, asbestos-cement piping, if

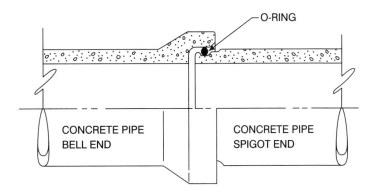

Figure 3.7 One example of a concrete pipe joint.
Figure 4.2.11.5, *National Standard Plumbing Code*, 1993 Edition. Reprinted with permission from the National Association of Plumbing Heating Cooling Contractors.

discovered in the field, should be handled and disposed of only by qualified and certified abatement professionals.

Glass piping used to be extremely popular for acid waste piping, particularly in aboveground applications. In fact, you still see it used in some pharmaceutical plants and laboratories for either acid waste piping or high-purity product piping. Joints are made with a stainless steel coupling that contains a special acid-resisting gasket.

Stone-based piping has many advantages. For the most part, except for the glass piping, it is relatively inexpensive, especially for the really large sizes of piping. These materials also have excellent strength characteristics and are suitable for high stress applications, such as under driveways, if properly installed. Additionally, these materials resist corrosion with most types of soils, even the more corrosive soils.

The disadvantages are apparent. Most of these materials are very heavy, thus increasing the installation cost. The majority of these materials are only manufactured in 4' or 5' lengths, thus further increasing the installation labor. Finally, they do not possess the corrosion resistance of the plastics, making them less suitable for systems that contain corrosive wastes. In recent years, more and more systems are changing over to one of the plastic systems, and, I suspect, this trend will continue in the future.

DRAIN, WASTE, AND VENT FITTINGS

There are a couple of important points to discuss regarding the fittings used on drain, waste, and vent piping.

First, most codes require that a certain radius be maintained when changing direction from the vertical to the horizontal direction or vice versa. For example, the *National Standard Code,* in Appendix F, requires the types of no-hub fittings shown in Figure 3–8 to produce a change in direction.

Maintaining these requirements for changes in direction will allow the wastes in the piping to flow in an even, controlled manner. Installing the piping with a turn that is too sharp will result in a system that constantly plugs and requires a high degree of maintenance. This is great news if you happen to own a plumbing repair company; it is bad news if you own the building.

Also keep in mind that the requirements for a change in direction apply only to drain, waste, and vent systems. These systems rely on gravity to move the waste along ("It all flows downhill"). Thus, these same requirements would not apply to the domestic water system as the domestic water system relies upon a change in pressure to move the water, not gravity.

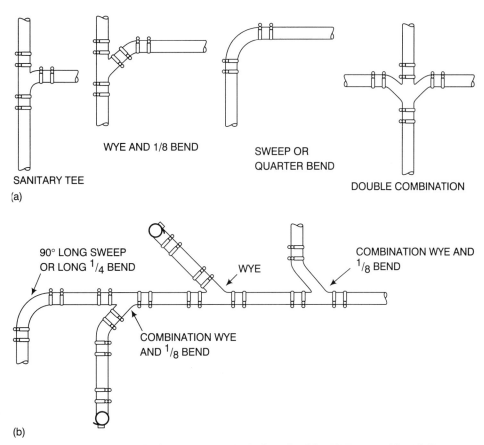

Figure 3–8 (a) Typical hubless fittings which are used for horizontal to vertical change of direction—elevation views shown. (b) Typical hubless fittings for horizontal to horizontal changes in direction. (c) Hydraulic jump examined. (d) Uses of short pattern fittings. Figures 2.3.3a, 2.3.4, 2.3.5b, and 2.3.6, *National Standard Plumbing Code,* 1993 Edition. Reprinted with permission from the National Association of Plumbing Heating Cooling Contractors.

The second main point to keep in mind is that a drain, waste, or vent fitting must not produce a joint where a ledge or other obstruction within the pipe is produced. This is to keep the solid wastes from "hanging up" on the ledge and thus allowing the pipe to plug. Therefore, this means that there is a difference between a copper DWV elbow and a copper elbow used on a domestic water system (sometimes called a *pressure elbow).* The same point holds true for plastic pipe and fittings.

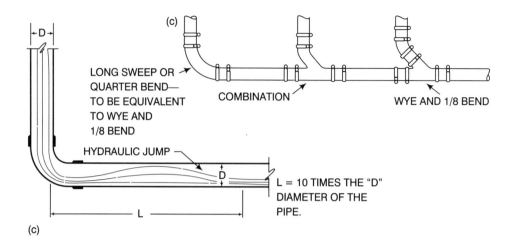

(c)

NOTE: LONG SWEEP FITTINGS ARE USED AT THE BASE OF STACKS TO MINIMIZE THE EFFECTS OF "HYDRAULIC JUMP." HYDRAULIC JUMP OCCURS WHEN THE HIGH VELOCITY FLOW FROM THE STACK ENTERS THE HORIZONTAL PIPING. THE VELOCITY OF THE FLUIDS CHANGE FROM APPROXIMATELY 15 FEET PER SECOND TO A LESSER VALUE. THE FLUID RISES AT A POINT WHICH VARIES TO TEN TIMES THE DIAMETER ("D × 10") OF THE PIPE DOWNSTREAM FROM THE DIRECTION CHANGE. THIS CONDITION CREATES EXCESSIVE PRESSURES IN THE AREA OF THE TURN, BUT THE PRESSURE EFFECTS MAY BE REDUCED THROUGH THE USE OF LONG PATTERN FITTINGS OR BY INCREASING THE DIAMETER OF THE BASE FITTING BY ONE DIAMETER. IT IS RECOMMENDED NOT TO INSTALL ANY BRANCHES IN THE "D × 10" ZONE UNLESS ADDITIONAL VENTING IS PROVIDED.

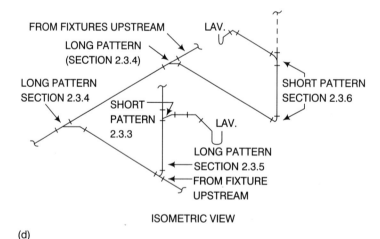

ISOMETRIC VIEW

(d)

NOTE: SHORT PATTERN DRAINAGE FITTINGS ARE SUITABLE FOR ANY INSTALLATION WHERE INTERSECTING DRAINS DO NOT MEET OR WHERE NOT MORE THAN ONE FIXTURE IS ON THE LINE.

Figure 3–8 *continued*

SUPPORTS AND HANGERS

Clearly, it is very important that all DWV piping is properly supported throughout its length. The need to keep the proper slope and to avoid sags and obstructions in the piping is paramount if the system is to work properly without a lot of unnecessary maintenance.

For underground DWV piping, the base of the trench should be graded to the appropriate slope and should be free of any rocks, stones, frozen dirt, or other unwanted material that may harm the piping during the backfilling and compaction process. The bottom of the trench should provide full support under the barrel of the pipe. Usually, this requires a small "bell hole" to be dug out under each pipe hub so that the trench bottom can support the barrel of the pipe (see Figure 3–9).

If the bottom of the trench contains unsuitable material, it will be necessary to overexcavate the trench bottom, remove the unsuitable material, and rebuild the bottom of the trench to the appropriate grade with gravel, crushed rock, sand, or other suitable material. It should be noted that some local plumbing codes and some pipe manufacturers require certain materials to have a base of gravel, crushed rock, sand, or other similar material. Further, these codes often require the use of gravel, crushed rock, or sand for the first 12" of backfill above the DWV line. PVC plastic, PE plastic, and fiberglass piping are examples of material that often have this type of additional requirement. In any event, extreme care should be used in backfilling and compacting the first 12" above any DWV line, regardless of the material, to help prevent cracking or damaging the piping.

DWV piping installed above the ground also needs to be carefully supported. There are a number of different types of hangers available for a wide variety of applications (see Figure 3–10).

Typically, the hangers should always be supported from the building structure. Do not support piping hangers from other piping, ductwork, or other similar devices, no matter how convenient they may be.

The maximum spacing between supports varies from code to code and should be thoroughly investigated by the installing contractor. From the ANSI A40-1993 Code, for example, the following support distances apply:

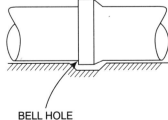

Figure 3–9 A bell hole.

Figure 3–10 Typical pipe supports.
Figure 1.2.58, *National Standard Plumbing Code,* 1993 Edition. Reprinted with permission from the National Association of Plumbing Heating Cooling Contractors.

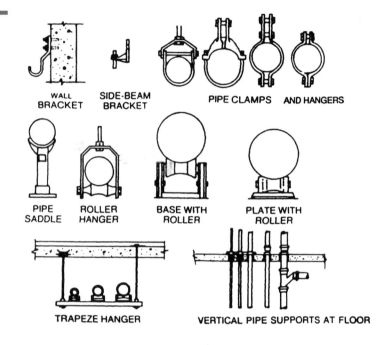

Table 3–1 Pipe Support Distances

	Vertical Spacing	Horizontal Spacing
Cast-iron Soil Pipe		
Lead joints	Each floor[1]	5'
No-hub	Each floor[1]	Each joint[2]
Copper Pipe		
1¼" or less	Each floor[3]	6'
1½" or more	Each floor[3]	10'
Steel Pipe		
Less than 1"	Each floor[4]	8'
1" or 1¼"	Each floor[4]	10'
1½" or more	Each floor[4]	12'
Plastic Pipe		
Rigid	Each floor[5]	3'
DWV	Each floor[5]	4'
Flexible	NA	32"

[1] Not to exceed 15'
[2] Not to exceed 4'
[3] Not to exceed 10'
[4] Not to exceed 25'
[5] Not to exceed 10' with a mid-story guide

Materials 37

VALVES

It is occasionally necessary either to install a device into the plumbing system that isolates that part of the system from the whole or to prevent flow in one direction in that part of the system. This is accomplished with valves. With drain, waste, and vent systems, it is always a hallmark of a good design to avoid all valves. The reason should be obvious. A valve, because of the way it is constructed, will inevitably provide a handy place for wastes to form an obstruction. Therefore, any valve in a waste or drain line should be viewed as a high maintenance condition which the owner will have to deal with over the life of the system.

This is not the case with domestic water systems. Not only are they conveying clear water, they are also pressurized systems; hence, the water flow is much more active and dependable.

The main types of valves used in plumbing systems are gate valves, ball valves, butterfly valves, and check valves (see Figures 3–11, 3–12, 3–13, and 3–14). Gate valves, ball valves, and butterfly valves are all used as isolation valves to isolate one part of the system from the main system or to isolate a

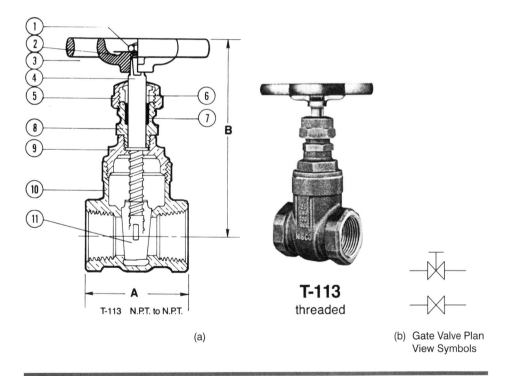

Figure 3–11 Gate valve.
Used with permission of NIBCO INC., Elkhart, Indiana.

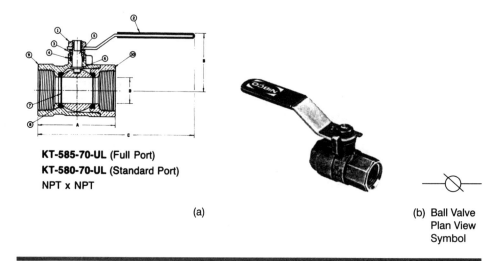

Figure 3–12 Ball valve.
Used with permission of NIBCO INC., Elkhart, Indiana.

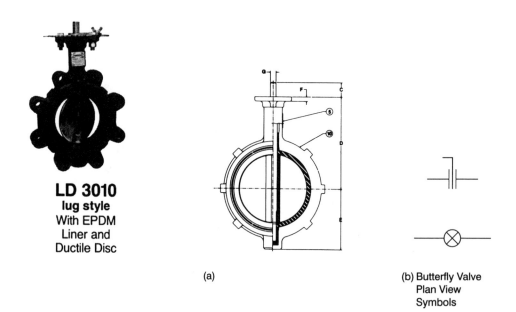

Figure 3–13 Butterfly valve.
Used with permission of NIBCO INC., Elkhart, Indiana.

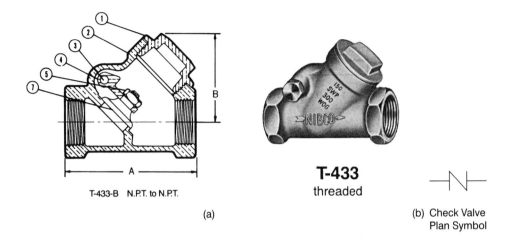

Figure 3–14 Check valve.
Used with permission of NIBCO INC., Elkhart, Indiana.

piece of equipment so that it can be serviced or replaced. The gate valve uses a gate to stop the flow within the pipe, whereas the ball valve uses a ball and the butterfly valve uses a disc. A gate valve should only be used for isolation duty: on and off applications. Because of the flow characteristics through a ball valve and a butterfly valve, these two valves can also be used to regulate flow within a pipe. This is sometimes useful, particularly in a domestic water system. A gate valve cannot accurately regulate flow, so it should not be used for that type of application.

A check valve has a hinged gate or a disc, depending on the type of check valve, that closes whenever the flow within the pipe reverses. This is particularly useful on the discharge of a water circulating pump, for example, when it is necessary to keep the pump from inadvertently running backwards when the pump is not activated.

REVIEW QUESTIONS

1. Why would you not want to use PVC plastic pipe and fittings in a return air plenum?
2. What is the purpose of the purple-colored primer on a PVC glued joint?
3. What pressure test is typically applied to drain, waste, and vent (DWV) systems?

4. Name at least three types of materials that you would consider for an application involving acidic wastes.
5. Name at least three types of materials that you would consider for domestic water piping, either above ground or below ground.
6. Name three methods of joining cast-iron soil pipe.
7. Why has lead piping been discontinued as a material for domestic water service?
8. What is the difference between copper piping with blue stenciled markings and copper pipe with red stenciled markings?
9. Why do some engineers prefer a brazed copper joint as opposed to a soldered copper joint?
10. What is the purpose of a check valve?
11. What types of fittings are required to change the direction of flow in a drainage system from vertical to horizontal? Why?
12. A drain, waste, and vent (DWV) fitting is different in what way from a similar fitting used in a nondrainage system?

CHAPTER 4

Sources of Water Supply and Points of Wastewater Disposal

WASTEWATER DISPOSAL

In almost all cases, the sanitary sewer drain line from a building connects to a city sanitary main, located outside the building, usually adjacent to, or in, the street. The city sanitary main, in turn, conveys the wastes to the municipal wastewater treatment plant for processing. Invariably, from the standpoint of public health as well as economics, connecting to a city main benefits both the community as a whole as well as the owner of the building.

There are situations, however, where there is simply no access to a city sanitary main. In these instances, alternatives must be considered. The most common alternatives are

1. Septic tanks with distribution fields
2. Aeration units
3. Stabilization ponds

Like all mechanical systems, each of these alternatives has certain advantages and disadvantages and should be installed with great care, as each of these alternatives increases the risk of contamination of the ground and surface water on the site. Accordingly, the presence of one of these systems on a site typically lowers the value of the land and building, thus reducing their marketability.

Septic Tank with Distribution Fields

A septic system is a very common and relatively simple system comprised of three components: the septic tank, the distribution box, and the distribution laterals (see Figure 4–1).

The purpose of the septic tank is to protect the distribution system from becoming clogged with solids. To accomplish this goal, the septic tank is designed to receive the various wastes from the building and allow the solids to settle to the bottom of the tank, where it is usually referred to as *sludge*. These solids are allowed to decompose by bacterial and natural processes. Typically, the tank also has a series of baffles or tees that retain those solids that float on the surface within the tank. These floating solids are usually

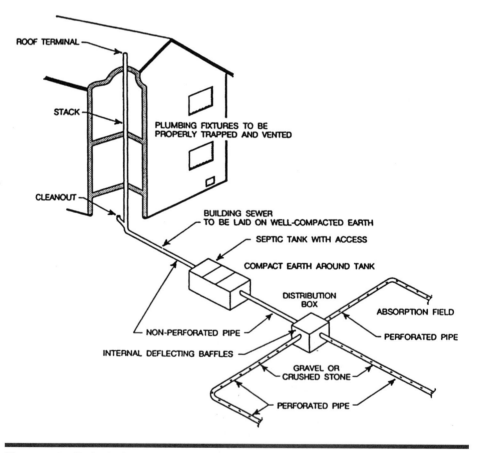

Figure 4–1 Typical private sewage disposal system.

Figure 1.2.42, *National Standard Plumbing Code,* 1993 Edition. Reprinted with permission from the National Association of Plumbing Heating Cooling Contractors.

referred to as *scum*. A properly installed, maintained, and operating septic tank should allow only wastewater, not solids, into the distribution system.

The purpose of the distribution system, or drainfield, is to dispose of the wastewater through absorption into the surrounding soil, through evaporation, through transpiration through the roots of grasses and plants above the drainfield, or, most commonly, by a combination of all three of these processes (see Figure 4–2).

Inevitably, one of the first questions that people ask about septic systems and drainfields concerns the odors and germs contained within the wastewater being absorbed by the soil. To answer the question, remember that most disease-causing germs have adapted to attacking man by becoming parasites. Therefore, these germs have become quite specialized in adapting to that particular environment. In a properly installed drainfield system (and the need for this cannot be stressed too much), these germs are evenly distributed throughout the soil, which is, of course, a completely different environment. These germs have a difficult time competing with other microorganisms that naturally live in the soil. For example, many of the common antibiotics that we use today were originally found and isolated from soil fungi, which are competing for the same nutrients in the soil as the germs from the wastewater. The germs from the wastewater are doomed to lose this battle.

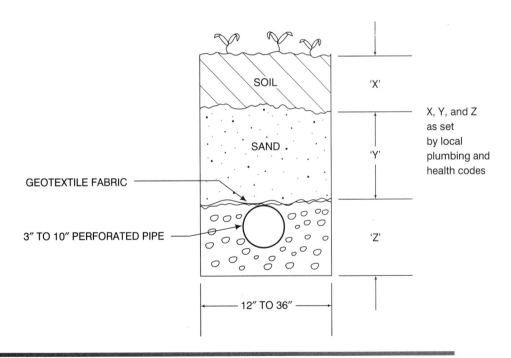

Figure 4–2 Elevation view of a typical lateral installation.

For this reason, the type of soil and its permeability is of prime concern when a septic system is being installed. The soil on any site should be carefully tested prior to the design of the septic system (and, hopefully, prior to the purchase of the land). A soil percolation test is one such test that measures the soil's ability to absorb water.

Typically, a soil percolation test consists of a hole or pit, 4" to 8" in diameter, dug to the bottom depth of the proposed absorption system. The sides and bottom of the pit should be carefully scraped with a sharp instrument to expose the natural soil. All loose material should be removed from the pit and the bottom of the pit covered with 2" of gravel or coarse sand. The hole is then filled with 12" of clear water, and the rate of fall of the water level is carefully measured. This percolation test then gives an approximate value for the soil's ability to absorb water.

The purpose of the distribution box is to ensure that the flow to each of the laterals in a distribution system or field is even. In a small system with only one lateral, a distribution box is not required.

Aeration Systems

An aeration unit differs significantly from a septic system in that an aeration unit uses nozzles to mix air throughout the sewage. This is necessary because the aeration unit uses a type of aerobic bacteria to decompose the sewage. The septic system, on the other hand, uses a type of anaerobic bacteria that does not require oxygen.

Typically, the sewage flows by gravity into a comminutor (or grinder) which discharges the sewage into the storage tank of the aeration unit. The air nozzles are located in this holding tank and constantly mix the sewage with air. After this mixing is complete, the sewage flows into a settling tank where the heavier residue is allowed to settle. Some aeration units may have more than one settling tank. Finally, the clarified liquid is brought into contact with a chlorination unit, before it leaves the aeration unit, to ensure that all harmful bacteria are completely neutralized.

A properly designed aeration unit that employs a sufficient amount of mixing, oxygen, and retention will completely reduce the sewage into a harmless mixture of carbon dioxide, water, and a residue. In fact, the water leaving an aeration unit is usually clarified enough that it can be dumped into any natural, free-flowing waterway. Obviously, this is rarely done due to liability concerns in the event that the aeration unit malfunctions. However, I recently became aware of a project in Texas where the discharge from the aeration units was used to water a large tract of land that contained cedar trees and a public right-of-way. This use of clarified effluent is a very forward-thinking method of disposal, in my opinion, and should be encouraged. Many local authorities, in a further effort to reduce liability, require an aeration unit to discharge into a stabilization pond, which is discussed in the next section.

Aeration units are usually installed for larger projects, where a building or a series of buildings is too large to be handled by a septic system. Therefore, these units can often be found on large manufacturing plants, housing subdivisions, or even whole towns, where the degree of treatment or the removal of pollutants is beyond the capacity of a septic system. Although not as common as septic systems, they are obviously a very useful alternative if a public system is not available and a large project is involved.

Stabilization Ponds

A stabilization pond is a lagoon or pond specially constructed for receiving wastewater. As the wastewater enters the lagoon, the heavier solids naturally gravitate to the bottom, where they are eventually decomposed by bacterial action. The lighter solids, which are suspended in the water, are also decomposed by bacterial action. Within a properly functioning stabilization pond, there are dissolved nutrients, such as nitrogen and phosphorus, as well as plenty of carbon dioxide, the byproduct from the bacterial decomposition. These elements, along with sunlight, support a thriving algae population, which is why most stabilization ponds characteristically have a bright green color. The byproduct of the algae population is oxygen, which feeds the bacterial population, thus completing the cycle that decomposes the sewage. As a final note, this bright green color usually produces a very inviting scene. Therefore, all such stabilization ponds should be secured with fencing and signs that alert the unsuspecting as to the true nature of the pond.

It should also be noted that a stabilization pond is designed to capture all of the wastewater entering the lagoon. For this reason, it is necessary to design a stabilization pond so that the evaporation rate is maximized, as this is the only method to remove the water. Therefore, most stabilization ponds are cone shaped to maximize the surface area of the pond. Additionally, trees, shrubs, and aquatic grasses should be eliminated so that wind action across the top of the pond is not reduced. This not only increases the evaporation rate, it also considerably reduces the mosquito population.

The soil conditions on the site of a stabilization pond are a critical concern, as they are for septic systems. The stabilization pond must be well sealed so that contamination of the ground water is avoided. Most local authorities will prohibit the installation of a stabilization pond if the soil is too porous, usually defined as a soil percolation test of 45 minutes per inch or greater.

WATER SOURCES

As is true for wastewater disposal, there are two main types of domestic water sources: public and private. Whichever source is used, some caution should be exercised. It has been my experience that water quality varies significantly, even over fairly small geographic areas. Very "hard" water is not all that uncommon and can cause a number of unpleasant problems, including excessive cleaning and maintenance, as well as odor and coloration problems. Additionally, I have found areas within the country where the water supply was actually very corrosive. A prudent construction manager will obtain a water sample test, conducted by a certified environmental testing laboratory, before the design phase of construction begins.

Public Water Supply Systems

Connection to the public water system, when available, is not only required in most communities, it is undoubtedly the best choice. Community systems following current federal regulations provide a safe, consistent flow of potable water at a usable pressure.

The original source of water for most municipalities is either ground water pumped from an aquifer or surface water from a reservoir, lake, or river. Municipalities using surface water usually have greater problems dealing with pollution than those who are fortunate enough to have an adequate source of ground water. Because of the amount of chemicals and compounds used today in every industry, including agricultural industries, the run-off of these chemicals and pollutants into our reservoirs has become a serious problem. In addition, the problem isn't affecting just those municipalities that use reservoirs; an alarming number of reports show that agricultural chemicals are starting to leach into the ground water and aquifers that supply potable water to many towns and cities in the United States.

To deliver potable water to consumers, most municipal systems employ either booster pumps, elevated tanks, or a combination of both. These energy sources will be connected, in most cases, to a network of underground water piping. This network of piping will then serve the community. It should also be noted that the depth of the water service piping will vary, as this network should always be installed well below the frost line. This prevents the mains from freezing in winter if the flow in the main stops. For example, the public main in Lincoln, NE, will be buried at a much greater depth than the public mains in Austin, TX. As with most aspects of the plumbing system, minimum frost depths are found in local and national plumbing codes.

As discussed briefly in a previous chapter, domestic water moves in a pipe due to a pressure differential. To create this pressure differential, we must add energy to the water. However, it is also important that we do not add too much energy to the water or the pressure in the piping will exceed the pressure that most plumbing fixtures are designed to accommodate. For this reason, most codes prohibit domestic water pressures above 80 psi (gauge). A prudent construction manager will also obtain a pressure test of the public main serving the construction site.

Private Water Sources

If a public system is not available, the next best alternative is usually a private well, pumping from a suitable water source. Since the quality and expense of a private well system can vary greatly, this should be fully explored before the design begins. Actually, the investigation of water sources needs to be done before the land is purchased to avoid a really unpleasant (and expensive) discussion with the owner.

One of the more important criteria in placing the well, after ensuring an adequate supply of safe water, is the location of the well on the site. Clearly, the well needs to be a safe distance, usually 100′ or more, from any private or public sewage system, such as a septic tank or stabilization pond. Further, the well needs to be a specified distance, usually 50′ or more, from any lot line. Finally, the well should be easily accessible, to provide maintenance and repair as needed. Again, these site isolation distances are typically specified in the site planning and subdivision regulations, as well as both local and national plumbing codes.

REVIEW QUESTIONS

1. Under what conditions would you be more apt to use an aeration unit as opposed to a septic system?
2. Should a private sewage system or a private well water system be used if a public system is available? Why or why not?
3. What is the purpose of the distribution field, or drainfield, of a septic system?
4. Why is soil effective in decomposing the germs often found in wastewater?
5. Why do most stabilization ponds have a green color?
6. Why is the pressure of a water supply system a concern?
7. Why are water supply mains installed at a lower depth in the ground in some areas of the country than in others?

CHAPTER 5

Plumbing Fixtures: Quantities, Location, and Layout

The starting point in most plumbing designs revolves around what is loosely referred to as the *plumbing layout*. The layout of the plumbing system is nothing more than the completion of three distinct steps. These steps are

1. Deciding what type of plumbing fixtures to provide
2. Deciding how many plumbing fixtures to provide
3. Deciding where to place the plumbing fixtures selected

DETERMINING THE TYPE OF PLUMBING FIXTURE

There are a large number of different plumbing fixtures available for a wide range of applications. For example, a plumbing fixture catalog from one of the top five manufacturers has hundreds of pages and is at least 3″ thick. Although the number of different fixtures available is staggering, there are really only six basic types of plumbing fixtures: water closets, urinals, sinks, showers/bathtubs, drinking fountains, and drains. All other fixtures are only modifications on the original in order to meet some special need or to produce some special aesthetic effect.

Water Closets

The most common type of plumbing fixture and undoubtedly the most important is the water closet, also known as a toilet. It is specifically designed to accept and discharge human wastes, including solids. Water closets are usually subdivided according to where they are mounted (floor mounted and wall mounted) and to how they are flushed (tank type and flush valve type). A handicapped-accessible water closet, either floor or wall mounted, has a rim that is installed 17" to 19" above the finished floor (see Figure 5–1).

The water closet in your home is probably a floor mounted, tank type. These water closets are typically very economical and, with the new regulations on water usage, use only 1.5 gallons per flush. Depressing the tank lever releases the contents of the tank to scour and flush the bowl. Once the tank is emptied, a water supply valve tied to a float opens, allowing water into the tank. As the water level rises, the float rises until it reaches the proper water level in the tank, at which time the float forces the water valve shut.

Floor-mounted water closets, however, are more difficult to clean around. Their low water usage, although very conservation minded, often doesn't do a very good job of cleaning the bowl. Hence, they have the reputation (well-founded in my mind) of not being as durable and requiring more mainte-

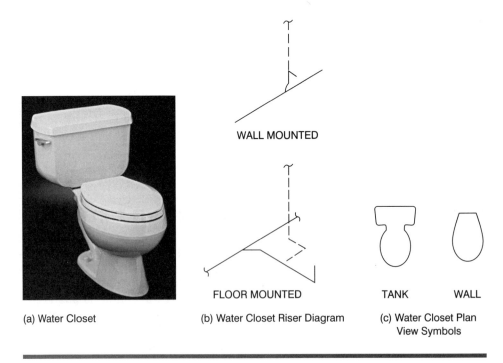

(a) Water Closet (b) Water Closet Riser Diagram (c) Water Closet Plan View Symbols

Figure 5–1 Water closets.

nance in a commercial application. Finally, because they are floor mounted, they obviously require a pipe directly underneath the fixture to carry away the wastes. This may create a headroom problem on the floor below. Because of their low cost, however, a floor-mounted, tank-type water closet is commonly found on both commercial and residential projects. Note, also, that the cost for a floor-mounted, tank-type water closet can exceed $1,400, depending on color, shape, and style. And the fixture isn't even made of gold.

Wall-mounted water closets, coupled with a flush valve, solve most of these problems. They are relatively easy to clean around, which is important if you have a lot of them in one location, as is often the case in a commercial application. The flush valve, in conjunction with either a siphon jet flushing action or a blowout flushing action, uses a lot more water at a higher pressure. Therefore, the cleaning action is significantly better, which, in turn, results in fewer maintenance and service problems. Besides the necessity of more water at a higher pressure, a wall-mounted water closet requires a special cast-iron support "chair" located in the wall behind the water closet (see Figure 5–2). The purpose of this chair is to support the full load of the fixture and the user. You do not want to attempt to have the wall support this load for the obvious reasons.

Figure 5–2 Closet carrier.
Reprinted courtesy of Kohler Company.

Unfortunately, the space for this chair is commonly forgotten by design professionals, which is another reason you oftentimes see a floor-mounted, flush-valve type of water closet installed as a compromise. Reprinted in Appendix C is the Plumbing and Drainage Institute's (PDI) excellent brochure which provides interior wall clearances for a wide variety of fixture supports.

Another relatively new development in the industry is the introduction of infrared sensing flush valves and faucets. When originally introduced, these devices were very expensive; however, they have since been steadily coming down in price. The infrared beam is set to sense an individual at a plumbing fixture, and, once that person steps away from the fixture, is programmed to automatically flush the fixture. This type of flush device greatly increases the overall sanitation of a public bathroom and should be encouraged where the budget allows.

Urinals

Urinals are another very important type of fixture. They are designed to accept and dispose of liquid human wastes only (see Figure 5–3).

Although most manufacturers still make a type of urinal that uses an elevated tank to supply water at the right pressure to flush the urinal, the vast majority of all urinals use a flush valve. Additionally, most urinals are of the wall-mounted type, as most local codes prohibit both floor-mounted models and the ever-popular trough urinal found in most sports stadiums, due to sanitation concerns. Fixture chairs are sometimes used for wall-type urinals, although not in many applications. They are typically used in those applications

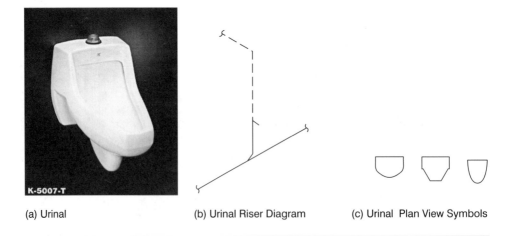

(a) Urinal (b) Urinal Riser Diagram (c) Urinal Plan View Symbols

Figure 5–3 Urinals.
Reprinted courtesy of Kohler Company.

where there is some concern that a person will actually lean on, or even sit on, the urinal. Accordingly, urinal chairs are often found in bars and junior/senior high schools, the only two institutions I know of where people have a tendency to support themselves by leaning, or even walking, on the urinal.

Sinks

Quite literally, this category contains everything including the kitchen sink. Sinks of all types are in wide use in a multitude of different applications. Surgeon's scrub sinks, service (or slop) sinks, lavatories, bar sinks, and scullery sinks are just a few of the more common ones used in plumbing systems. Sinks can be made of porcelain, stainless steel, plastic, fiberglass, or any other nonporous material. Sinks can be wall mounted, floor mounted, set into a countertop, or free standing with legs (see Figures 5–4 and 5–5).

Most sinks have a faucet mounted either on the rim of the sink or on the wall or countertop immediately adjacent to the sink. Some sinks, such as a flushing rim sink, can handle solids. Other sinks have multiple compartments, one of which may be equipped with a food waste grinder, which is another way of dealing with solids. One note of warning concerning service sinks or other janitorial sinks that have faucets with a threaded end. Many custodians attach a hose to these faucets for a variety of reasons. Any faucet with a threaded end should have a vacuum breaker either built into the faucet or installed immediately above the faucet to ensure that back-siphonage does not occur through the use of the hose. Additionally, the outlet of all faucets should be installed above the flood level rim of the fixture to ensure that used or contaminated water does not reenter the water supply by rising above the faucet outlet.

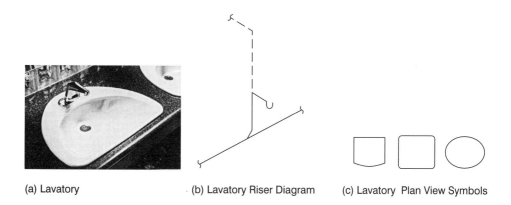

(a) Lavatory   (b) Lavatory Riser Diagram   (c) Lavatory Plan View Symbols

Figure 5–4 Lavatories.
Reprinted courtesy of Kohler Company.

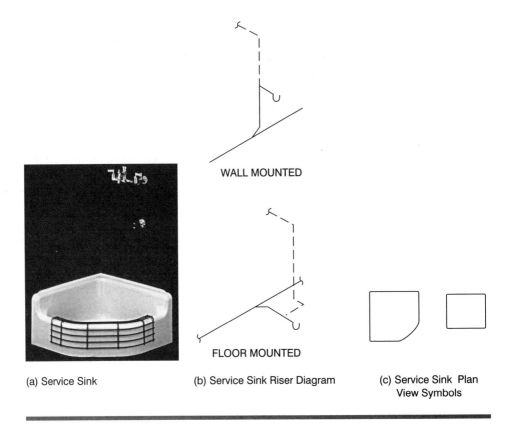

Figure 5–5 Service sinks.
Reprinted courtesy of Kohler Company.

Bathtubs and Showers

A wide variety of bathtubs and showers are available to provide bathing facilities for the occupants of a building. Bathtubs and shower can be used in a single combination fixture or can be completely separate. Bathtubs can be made of enameled steel, enameled cast iron, fiberglass, or acrylics and may include such popular options as whirlpools (see Figure 5–6).

Showers are equally diverse (see Figure 5–7). You can order pulsating shower heads, "low flow" water conservation heads, and handheld, handicapped-accessible shower heads, just to name a few. Additionally, group showers, sometimes called column showers, are popular on commercial or institutional projects where a large number of people must be served at one time.

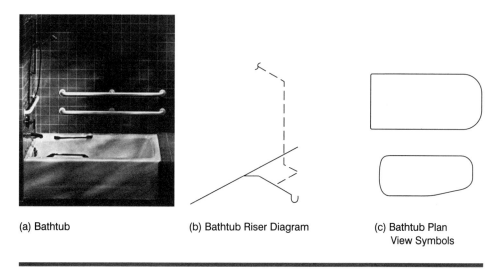

(a) Bathtub   (b) Bathtub Riser Diagram   (c) Bathtub Plan View Symbols

Figure 5–6 Bathtubs.
Reprinted courtesy of Kohler Company.

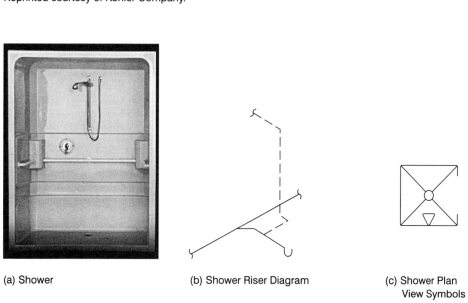

(a) Shower   (b) Shower Riser Diagram   (c) Shower Plan View Symbols

Figure 5–7 Showers.
Reprinted courtesy of Kohler Company.

Drinking Fountains

Drinking fountains have recently received more attention than they have in the past. Principally, this is due to the new handicapped accessibility rules required by the Americans with Disabilities Act (ADA) and by recent concerns with lead in the drinking water. Most, if not all, of the newer model drinking fountains are completely lead free in design, so that no lead leaches out of the fixture and into the water stream being consumed. Most drinking fountains installed today are of the handicapped-accessible type, whether or not they are installed in a handicapped-accessible area (see Figure 5–8).

Floor Drains, Roof Drains, and Cleanouts

Although drains and cleanouts may not be considered "fixtures," they are an important part of the plumbing system and perform much the same purpose as most other plumbing fixtures. They are usually considered together as all three are typically made of cast iron and, therefore, are usually purchased from the same vendor for an entire project.

Floor drains are used in a wide variety of applications. As drains set in the floor, they can provide a convenient point of disposal for overflow lines or

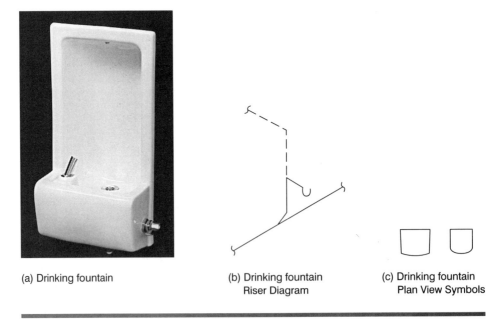

(a) Drinking fountain

(b) Drinking fountain Riser Diagram

(c) Drinking fountain Plan View Symbols

Figure 5–8 Drinking fountains.
Reprinted courtesy of Kohler Company.

relief lines, such as the discharge line from a safety relief valve. Additionally, they can aid in cleaning and maintenance; hence, floor drains are commonly found in larger bathrooms. In fact, some local codes now require a floor drain installed in every bathroom that contains two or more water closets. A "first cousin" to the floor drain, the floor sink is commonly used to accept the indirect discharge of commercial dishwashers, ice machines, and other similar appliances. The reason a floor sink is used in this application is that it allows the appliance to connect indirectly to the plumbing system through an air gap. Said another way, the discharge pipe from the appliance terminates approximately 2″ above the flood level rim of the drain. Therefore, if the drain and waste system does back up from the drain, it cannot enter an appliance, such as a dishwasher, where the wastewater may contaminate the clean dishes.

Roof drains, as the name implies, are drains set in the roof to accept rainwater, snowmelt, and other clear water wastes. As with floor drains, they come in a wide variety of sizes, shapes, and materials in order to meet a wide spectrum of applications. They also have a "first cousin," called an *area drain*. An area drain serves the same purpose as a roof drain, except that it is installed somewhere other than a roof, such as at the bottom of an exterior staircase that extends below grade.

Cleanouts are devices that allow access to the DWV system for the purpose of cleaning the piping or removing an existing blockage. They can be set into the floor or recessed into a wall, as necessary, in order to provide access to the DWV piping. Most plumbing firms use a cable (commonly referred to as a *snake)* with a blade or cutting edge on the front of the cable to clean or remove obstructions within DWV piping. The cable is attached to a motor which twists the cable through the piping, thus scouring the pipe clean. I mention the method of cleaning because it is very important that a cleanout be located so that there is enough room for maintenance workers and the cabling machine to perform their work. All too often, I have seen cleanouts installed in a location that makes them, for all practical purposes, useless. Most codes, including the A-40 code, require cleanouts in horizontal piping every 75′ for piping 4″ in diameter and smaller, and every 100′ for piping larger than 4″ in diameter. Additionally, cleanouts should be installed in the base of every vertical stack and at the end of a horizontal run. In reality, good common sense is the best tool that you can use in placing cleanouts.

Sewage Ejectors, Sump Pumps, and Booster Pumps

Pumps are another example of devices that are not often considered to be plumbing fixtures, yet serve a very important part within the plumbing system. A sewage ejector is a pump that has the capability to handle human wastes, including solids. A sump pump, on the other hand, has the capability to handle only clear water or liquid wastes. A booster pump is used on a domestic water system to boost or increase the pressure of the water.

The purpose of any pump is to impart energy into a fluid. For a sewage ejector or a sump pump, this energy is used to propel the fluid to a higher elevation (see Figure 5–9). This is very important in those situations where the lowest drain or waste line within a building is installed below the elevation of the city main which serves the building. In other words, the building drain cannot drain by gravity into the city main.

Clearly, sewage ejectors and sump pumps should be avoided, if at all possible. As we stated earlier, every mechanical device eventually fails, and sump pumps and sewage ejectors typically have higher than average maintenance costs. Additionally, the sumps, no matter how well sealed or vented, will have some odor migration. If a sewage ejector is necessary, most engineers recommend a duplex installation consisting of two pumps and a control panel to set off an alarm in the event the sewage rises above a set level in the basin. This is because most sewage ejectors fail during a high flow situation or during a

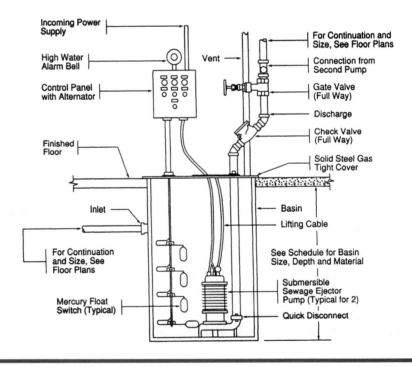

Figure 5–9 Sewage ejector.
Reprinted with permission of *Plumbing Engineer* magazine.

power outage. This obviously results in a backflow condition in those waste lines leading to the sewage ejector and, inevitably, the flooding of the room in which the sewage ejector is located. Common sense tells us that this is another one of those situations we really want to avoid.

Booster pumps are commonly required on the domestic water system in buildings of four stories or more, as the city main pressure is not usually sufficient to deliver the water to a plumbing fixture at its maximum required pressure on an upper floor. Booster pumps are commonly designed to have all bronze construction where the domestic water comes into contact with the construction of the pump to avoid discoloration of the water. As with the sewage ejectors, most engineers specify at least two pumps to a booster pump set, and even three or more pumps may be used. This not only provides much-needed redundancy, it also provides a much more efficient operation as the demand for domestic water varies. The newest trend in controlling booster pumps is to use a variable speed drive that varies the speed of the pump based upon the pressure of the most distant water supply pipe (see Figure 5–10).

Figure 5–10 Booster pump.
Reprinted with permission of ITT Bell & Gossett.

DETERMINING THE QUANTITY OF PLUMBING FIXTURES

Deciding how many plumbing fixtures to provide is not always the easiest thing to do. Most plumbing codes, the ANSI A-40 Code among them, requires a minimum number of plumbing fixtures in each building, based on the type of building and the number of people of each sex within that building. This information is found in Table 4 of the ANSI-A40 Code and is reproduced in Appendix C.

The problem with any code or law, of course, is that it only sets a minimum standard. It does not address, nor does it pretend to address, the specific needs of the individual owner. You just have to attend a basketball game or a football game and witness the long lines of rather anxious-looking people standing outside the bathrooms to grasp an understanding of what can happen if you only focus on the minimum standard. Added to this problem is the fact that many local plumbing codes have not been kept current with new developments and understandings within our industry. For instance, many older codes seriously understate the minimum number of plumbing fixtures required for females because these older codes, some dating back to the 1940's, fail to take into account the recent influx of females into the workforce. An unwary construction manager who relies on the code may therefore be led into producing an embarrassingly inadequate amount of plumbing fixtures. It is always well to remember that any code is nothing more than a listing of the *minimum* standards that must be achieved. In all reality, this statement ought to be considered another one of the "great truths" of plumbing.

The best course of action in determining the number of plumbing fixtures required is to work closely with the architect and the owner and come to a consensus on what the needs of the owner really are and how best to meet those needs. Once this is done, the astute construction manager will double check this consensus with the applicable code to make sure that all of the minimum provisions of the code have been met.

DECIDING WHERE TO PLACE THE PLUMBING FIXTURES

The placement of the plumbing fixtures within the building should always be the product of the architect and the owner, with input from the mechanical engineer. It is very important that the plumbing fixture location is compatible with the flow of the people within the building. Improperly placed bathrooms and other plumbing fixtures (particularly drinking fountains) will lower productivity, increase maintenance, and create a number of logistics problems for the owner.

The actual position of the plumbing fixture within the space is also very important. Again, the ANSI-A40 Code contains minimum required clearances for various plumbing fixtures. These clearances define the amount of spacing between plumbing fixtures or between plumbing fixtures and walls or partitions. These dimensions are found in Figure 5–11.

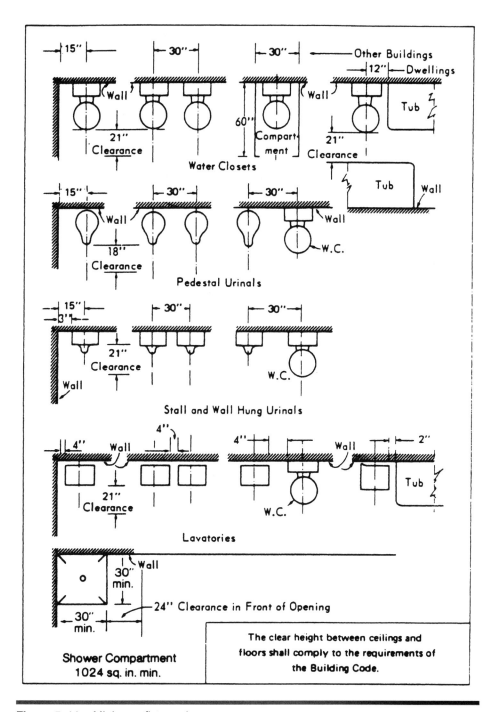

Figure 5–11 Minimum fixture clearances.
Figure 1, ANSI A40-1993 Standard, *Safety Requirements for Plumbing,* p. 172. Reprinted with permission.

Recently, the federal government injected a "wild card" into the process of determining the proper fixture layout. This wild card is called the Americans with Disabilities Act (ADA) and, in effect, produces the new design standard for plumbing fixture layouts. These new dimensional standards are also reprinted in Appendix C and should be carefully studied and applied. The reason many designers consider the ADA a "wild card" is that the ADA is vague on when and how the law is to be implemented in an existing structure. The ADA requires implementation if the alterations are "readily achievable," and many owners, as well as designers, have a lot of trouble defining the term "readily achievable" for their particular application. It should be stressed, however, that an improper application of the ADA regulations could leave you and your client liable for a painful and protracted discrimination suit.

One final note on plumbing layouts needs to be stressed. It is very important to remember that virtually all plumbing fixtures have connections to the piping from either below the fixture, from behind the fixture, or both. Some plumbing fixtures, such as wall-mounted water closets, also require a chair support located in the wall behind the fixture. The importance of providing plenty of space within the wall behind the fixtures, and above the ceiling for the piping, becomes apparent. Unfortunately, apparent as it is, this is still a common problem on many construction sites. For example, a 4" cast-iron waste stack will not fit inside a 4" wall, as the diameter of the hub is greater than 4". Also, the 4" wall designation typically refers to an "outside to outside" dimension, which, clearly, will not contain the 4" pipe. This problem is often accentuated by the fact that many designers attempt to stack the bathrooms one on top of another on multi-story buildings or position the bathrooms back-to-back if the bathrooms are on the same story. This type of layout produces a much more economical design in that the drain, waste, vent, and water piping for both sets of plumbing fixtures can be located in the same wall. Often, the designers will then refer to a wall that contains plumbing piping as a *wet wall*. As these wet walls are a point where several different trades converge, they require a great deal more coordination by all parties involved.

REVIEW QUESTIONS

1. What special device is required by a wall-mounted water closet? Why is this device required?
2. What is the definition of a wet wall?
3. What is meant by the term "back-to-back" bathrooms?
4. What is the minimum interior wall dimension required for back-to-back bathrooms using block wall construction and wall-mounted water closets?

5. What is the minimum and maximum rim height off the finished floor for a handicapped-accessible water closet?
6. What is the minimum center-to-center dimension between two water closets?
7. For a two-story office building employing 40 women and 40 men, what is the *minimum* number of water closets, lavatories, urinals, drinking fountains, and slop (service) sinks required?
8. What is the *standard* width of a handicapped-accessible water closet stall?

CHAPTER 6

Sizing the Drain, Waste, and Vent System

Sizing the piping within the drain, waste, and vent system is not that complicated. As with any process, there are a few basic rules that need to be followed. These rules follow the fundamentals that we learned in previous chapters. Before we apply these fundamentals to sizing, however, we need to understand two important tools used in sizing drain, waste, and vent systems: drainage fixture units and riser diagrams.

DRAINAGE FIXTURE UNITS

One important concept that needs to be addressed is how we measure the amount of waste that discharges from a particular plumbing fixture. We also have to ask ourselves how much water the same fixture will use. It is important to note that these two values are not always the same. The amount of waste leaving a plumbing fixture does not always equal the amount of water connected to the plumbing fixture.

All plumbing fixtures are given a rating based on the amount of waste leaving the fixture and the amount of water used by the fixture. For drainage systems, this rating is called the drainage fixture unit (DFU). The DFU of a plumbing fixture is based on three factors:

1. The volume of discharge contained in a single use (i.e., the amount of discharge)
2. The duration of a single discharge (i.e., how long that fixture drains)
3. The average time between successive operations of the fixture.

For the water supply to the fixture, this rating is called the water supply fixture unit (WSFU). The WSFU is also based on three factors:

1. The volume of water used in a single operation of the plumbing fixture
2. The duration of a single use of a plumbing fixture
3. The average time between successive operations of the plumbing fixture

Both the DFU and the WSFU are nothing more than a probability. It should be stressed that the DFU and WSFU figures are *not* the same and, thus, are *not* interchangeable. This is a common mistake. The DFU value for various plumbing fixtures is presented in Table 2–2, in Chapter 2.

It is interesting to note that this system of rating was first developed in the early 1940's by Professor Roy Hunter. In a later chapter, you will have the opportunity to use the curves derived from Hunter's research (appropriately enough, these are called "Hunter Curves") to size the water supply piping. Most codes, and, therefore, most designers, use basically the same system as originally developed by Hunter. The problem with this is that many things have changed since these tables and ratings were first developed. For example, a new generation of plumbing fixtures have been developed which are designed to conserve water. These new "low flow" water closets use only 1.6 gallons per flush as opposed to a standard water closet that uses 3.0 gallons per flush. Therefore, some of the DFU values may well be overstated for these types of fixtures.

Additionally, the demographics of the workplace have changed dramatically since World War II. These changes also need to be addressed in the code if an effective and satisfactory plumbing design is to be produced. Currently, the industry is researching these and other issues in order to develop a better system of rating the probability of flow to and from plumbing fixtures. At this time, however, this research has not produced a system materially better than the original DFU system developed by Hunter.

RISER DIAGRAMS

The riser diagram is one of the best tools that you can use to analyze and solve plumbing problems. Additionally, virtually every local building code requires a drain, waste, and vent riser diagram to be shown on the plumbing drawings, prior to the issuance of a building permit or a plumbing permit. Therefore, learning how to draft and read a riser diagram is extremely important to every construction manager.

A plumbing riser diagram is nothing more than a three-dimensional rendition of the drain, waste, and vent piping, or the domestic water piping, for a number of plumbing fixtures. Typically, separate riser diagrams are produced for the DWV system and for the domestic water system. However, for very small bathrooms or systems, the diagrams are sometimes combined. Riser diagrams are usually done for each separate bathroom or riser, unless the building is quite small. In that case, a single riser diagram may cover the entire building.

The rules for drawing a riser diagram are simple. They are as follows:

1. Obtain or develop a plan view perspective of the bathrooms or plumbing fixtures.
2. Connect all plumbing fixtures with drain, waste, and vent piping using lines on 30/60/90° angles to achieve a three-dimensional perspective.
3. Drain and waste piping is shown as a solid line.
4. Vent piping is shown as a dashed line.
5. Every plumbing fixture must have a trap. (Exception 1: remember that some plumbing fixtures have a built-in, or integral, trap that is not shown on a riser diagram.)
6. Every trap must have a vent. (Exception 2: remember that some types of vents provide venting to more than one trap at a time.)
7. Every plumbing fixture must be labeled or identified.
8. All piping must be properly sized and the size indicated on the drawing. The term "properly sized" means the piping must not be oversized or undersized.
9. Cleanouts must be shown for the proper maintenance of the system.
10. Riser diagrams are rarely drawn to scale.

As with any other discipline, drawing riser diagrams becomes much easier after you have completed a few on your own (see Figures 6–1 and 6–2).

Figure 6—1 Line diagram #5.

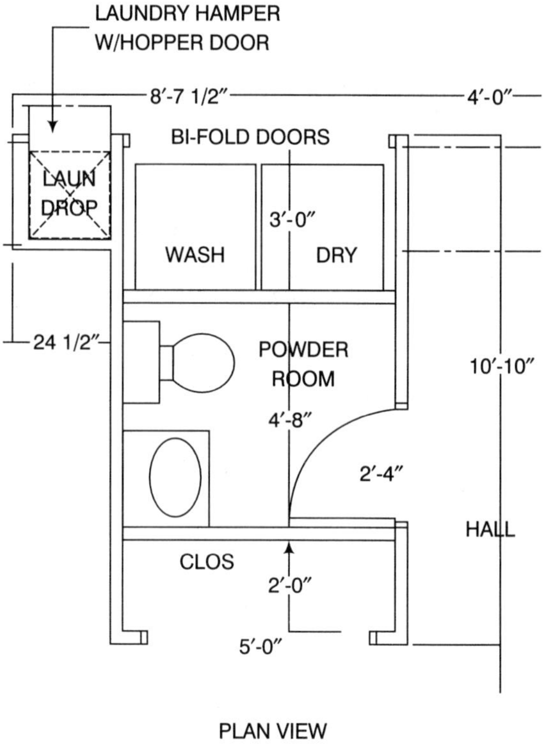

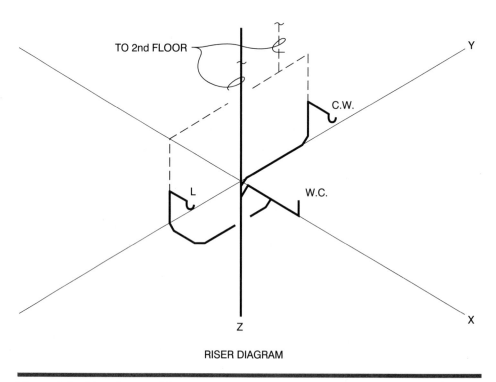

Figure 6–2 Line diagram #6.

SIZING THE DRAIN AND WASTE SYSTEM

We are finally ready to learn how a simple drain and waste system is sized. It is really not difficult at all. The steps are as follows:

1. Determine the number of plumbing fixtures to install As discussed earlier, this determination is made by the owner, architect, and mechanical engineer.

2. Lay out the plumbing fixtures in the space This is also generally the product of the owner, architect, and mechanical engineer, with guidance provided by federal, state, and local codes (do not forget about the ADA!).

3. Determine the location of the building on the site and the location and depth of the city sewer main This will allow you to determine the direction of the building drain and the maximum slope of the lowest drain line. It may also tell you that the city main is at too high an elevation to allow your building wastes to drain by gravity into the city main. In this situation, you would have to drain the building wastes into a sewage pit or sump and then pump the sewage to a higher elevation, where it could then flow by gravity into the city main.

Most designers like to keep the sanitary sewers to a maximum depth of 8' to 10', as this is the maximum working depth of a standard backhoe. If your sewer is deeper than this, your installation costs will skyrocket.

Finally, keep in mind that most codes specify a minimum slope or grade for all piping. For the ANSI-A40 Code, the slope must not be less than ¼" per foot for waste pipes 3" and smaller or less than ⅛" per foot for waste pipes 4" and larger.

4. Generate a simple isometric riser diagram, showing the plumbing fixtures and the waste and vent piping.

5. Determine the minimum trap size for each individual plumbing fixture from Table 2–2, in Chapter 2.

6. Determine the drainage fixture unit (DFU) rating for each plumbing fixture, also found in Table 2–2.

7. Add the DFU values for each branch and section, from the farthest end of the system toward the point where the building drain leaves the building.

Most codes, the ANSI-A40 Code among them, size the building drain and sewer differently than they size the horizontal branches and stacks. Accordingly, one table is used for sizing the building drain and sewer, and a different table is used for sizing horizontal branches and stacks. For the ANSI-A40 Code, Table 6–1 is used for the building drains and sewers, and Table 6–2 is used for horizontal fixture branches and stacks.

Table 6–1 Maximum loads that may be connected to building drains, branches of the building drain, and building sewers.

Diameter of pipe, in.	Fall per foot			
	1/16 in.	1/8 in.	1/4 in.	1/2 in.
	DFU Values			
2			21	26
2½			24	31
3			42*	50*
4		180	216	350
5		390	480	575
6		700	840	1,000
8	1,400	1,600	1,920	2,300
10	2,500	2,900	3,500	4,200
12	2,900	4,600	5,600	6,700
15	7,000	8,300	10,000	12,000

*Not over two water closets or two bathroom groups.
Table 17, ANSI A40-1993 Standard, *Safety Requirements for Plumbing.*
Reprinted with permission

There are a few rules to keep in mind when sizing the drain and waste piping (see Figure 6–3). These simple rules are as follows:

1. The stack cannot be smaller than the largest horizontal branch connecting to the stack. For example, a 4″ horizontal branch may not connect to a 3″ vertical stack.
2. Some codes require that each building have one waste stack that is the same size as the building drain, or at least 3″ in size.
3. No horizontal waste pipe can be smaller than the pipe that precedes it. Said another way, a horizontal waste pipe cannot be reduced in the direction of flow. A horizontal waste pipe can be reduced, away from the flow, as the waste pipe collects fewer plumbing fixtures.
4. An above ground waste or drain pipe cannot be less than 1¼″ in diameter.
5. A below ground waste or drain pipe cannot be less than 2″ in diameter.

Table 6–2 Maximum loads that may be connected to horizontal fixture branches and stacks.

Diameter of Pipe, in.	Sizing for Any Horizontal Fixture Branch[a]	Total Fixture Units Allowed on One Stack of Three-Branch Intervals or Less	Stacks with More Than Three-Branch Intervals	
			Total for Stack	Total at One-Branch Interval
	DFU	DFU	DFU	DFU
1½	3	4	8	2
2	6	10	24	6
2½	12	20	42	9
3	20[b]	48[b]	72[b]	20[b]
4	160	240	500	90
5	360	540	1,100	200
6	620	960	1,900	350
8	1,400	2,200	3,600	600
10	2,500	3,800	5,600	1,000
12	3,900	6,000	8,400	1,500
15	7,000			

[a]Does not include branches of the building drain.
[b]Not more than two water closets or bathroom groups within each branch interval nor more than six water closets or bathgroups on the stack. Stacks shall be sized according to the total accumulated connected load at each story or branch interval and may be reduced in size as this load decreases to a minimum diameter of 0.5 of the largest size required.
Table 18, ANSI A40-1993 Standard, *Safety Requirements for Plumbing.* Reprinted with permission.

6. A 1¼" drain pipe, if used, can handle only 1 DFU.
7. A drain line cannot be smaller than the trap arm to which it is connected.
8. Some codes allow a 3" trap arm for a floor-mounted water closet; however, if at all possible, always use a 4" trap arm. The additional cost is minimal, and the savings in maintenance is significant.

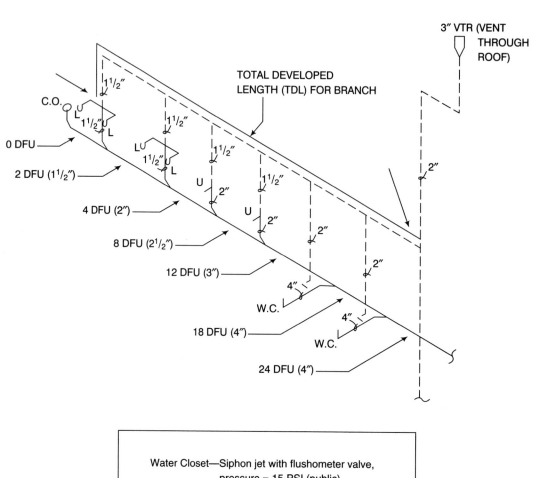

Figure 6–3 DWV sizing example.

SIZING OF THE VENT SYSTEM

The proper sizing of the vent system is really no more difficult than the sizing of the sanitary waste system. It is important to remember that the amount of air required by the plumbing system is proportional to the amount of waste flowing in the drainage pipe. Therefore, we use the same DFU value to size the vent piping that we used to size the waste piping.

The steps in properly sizing a vent stack are as follows:

1. To size a vent stack, first determine the total number of DFU's that connect to that stack.
2. Determine the total developed length (TDL) of the vent stack, from where it connects to the waste stack to where it terminates in the atmosphere. The TDL is the length of the pipeline measured along the centerline of the pipe and fittings. Do not forget to add some additional length to account for the fittings (elbows, tees, and so on) found in the vent stack that will increase the friction loss. Often, many engineers will add 15% to 25% to the actual length, depending on the complexity of the system, to account for the fitting loss.
3. Determine the size of the waste or soil stack, as discussed previously.
4. Using the values obtained in Items 1, 2, and 3 and using Table 6–3 from the ANSI-A40 Code, determine the size of the vent stack. Do not forget that, by definition, a vent stack must be installed *full size* for its entire length.

To size the branch venting, we use Table 6–4, just as we did for stack and vent stacks. Because the venting requirements for vent stacks and stack vents are more severe than that for horizontal and branch vents, we actually end up introducing a small and reasonable factor of safety into our calculations by using Table 6–4. The steps used in sizing branch vents are as follows:

1. To size the branch vent, determine the TDL from where the individual vent connects to the trap arm to where the branch vent connects to either the vent stack or stack vent.
2. Determine the total DFU value for all plumbing fixtures connected to that section of branch vent.
3. Determine the size of the soil or waste pipe, as discussed previously.
4. Using the values obtained in Items 1, 2, and 3 and Table 6–4 from the ANSI-A40 Code, determine the size of the branch vent.

Table 6–3 Size of circuit or loop vent.

Soil or waste pipe diam. (in.)	Fixture units (max. number)	Diameter of circuit or loop vent (in.)				
		2	2½	3	4	5
		Max. horizontal length, ft				
3	10	20	40	100	—	—
3	30	—	40	100	—	—
3	60	—	16	80	—	—
4	100	7	20	53	200	—
4	200	6	18	50	180	—
4	500	—	14	36	140	—
5	200	—	—	16	70	200
5	1100	—	—	10	40	140

Table 23, ANSI A40-1993 Standard, *Safety Requirements for Plumbing.* Reprinted with permission.

As with the sanitary and waste sizing, there are a number of simple rules to keep in mind. They are as follows:

1. An above ground vent line cannot be less than 1¼" in diameter.
2. A 1¼" vent, if used, can handle only 1 DFU.
3. An underground vent line cannot be less than 2" in diameter.
4. An individual vent must be within a certain distance of the trap which it serves. This information is found in Table 2–3 of the ANSI-A40 Code, in Chapter 2.
5. An individual vent for a plumbing fixture must not be less than one-half the size of the drain to which it connects.
6. An individual vent should not be larger than the drain to which it connects.
7. A downstream section of vent piping can never be smaller than the upstream section of vent piping to which it connects.
8. The horizontal section of a branch vent must slope toward a drain.
9. A vent line must extend at least 6" above the flood level rim of the fixture before it can be turned to the horizontal.

Table 6–4 Minimum diameters and maximum lengths of vent stacks and stack vents.

Size of soil or waste stack, in.	Fixture units connected	Minimum diameter of vent required, in.								
		1¼	1½	2	2½	3	4	5	6	8
		Maximum length of vent, ft.								
1½	8	50	150							
2	12	30	75	200						
2	20	26	50	150						
2½	42		30	100	300					
3	10		30	100	100	600				
3	30			60	200	500				
3	60			50	80	400				
4	100			35	100	260	1000			
4	200			20	90	250	900			
4	500			20	70	180	700			
5	200				35	80	350	1000		
5	500				30	70	300	900		
5	1100				20	50	200	700		
6	350				20	50	200	400	1300	
6	620				25	50	125	300	1100	
6	960				15	24	100	250	1000	

Table 24, ANSI A40-1993 Standard, *Safety Requirements for Plumbing*. Reprinted with permission.

76 Chapter 6

REVIEW QUESTIONS

1. What is the difference between the building drain and the building sewer?
2. What is wrong with the sizing of this horizontal waste line?

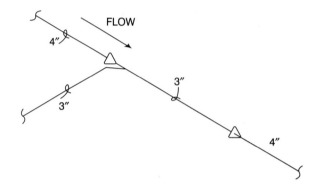

3. What is the smallest waste line allowed under a slab on grade?
4. Can a vent stack be reduced in size? Why or why not?
5. Name three things wrong with this riser diagram.

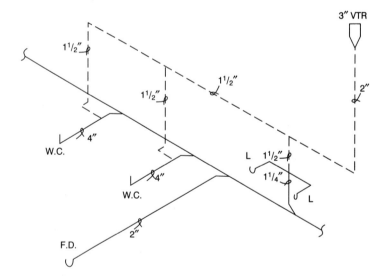

Sizing the Drain, Waste, and Vent System 77

6. Draw a riser diagram for the following bathroom. Size all piping.

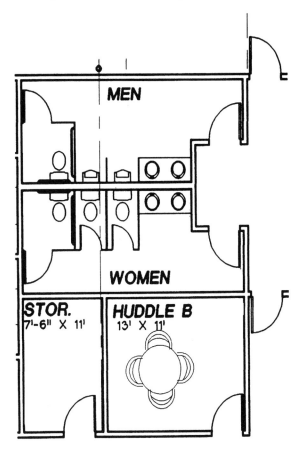

7. Draw a riser diagram for the following bathroom group. Size all piping.

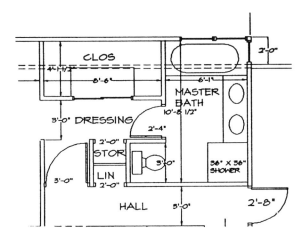

8. Draw a riser diagram for the following diagram. Size all piping.

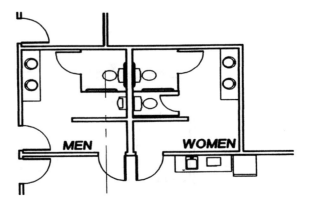

CHAPTER 7

Storm Water Systems

INTRODUCTION

Storm systems, sometimes referred to as rainwater systems, are another important component of the plumbing system within a building. A properly designed storm drainage system will accomplish the following goals:

1. Prevent the roof system of the building from overloading due to the weight of the water collecting on the roof
2. Prevent water from collecting around the foundation of the building and thus entering the building
3. Protects the land, and your neighbor's land, from damage due to erosion

These are terribly important goals to keep in mind. To lend some perspective to the importance of these goals, keep in mind that over 27,000 gallons of water will fall on an acre of ground during a 1" rainfall. The weight of this 27,000 gallons of water is 225,342 pounds. It is not surprising, then, that a malfunctioning roof drain system that allows water to accumulate on a roof is always viewed as a very serious problem.

FUNDAMENTALS

A quick discussion on combination sanitary/storm systems is probably in order at this point. Most codes, and even most texts on plumbing design, mention or allow the installation of piping that will simultaneously carry sanitary wastes and storm wastes. In fact, these types of systems used to be quite common, and there is a good chance that you will encounter some of these systems in your career as a construction manager.

However, due to increasingly strict government regulations on the treatment of sanitary wastes, most local governments strictly prohibit any connection between the sanitary system and the storm system. This only makes good sense. With the new government regulations on the treatment of sanitary wastes, it costs a local government a great deal of money to treat and dispose of sanitary wastes. By mixing in rainwater, a liquid that needs no treatment whatsoever, a local government is drastically increasing its treatment costs, to the detriment of the taxpayer. In fact, most local governments now have rather severe fines and penalties for dumping rainwater into the sanitary sewer. Unfortunately, this scenario frequently occurs when a homeowner decides to discharge their footing drain sump pump (which contains only ground water) into the nearest available sanitary sewer. It is my opinion that all storm water systems should always be kept separate from the sanitary systems. Therefore, I have chosen not to include the sizing tables and data on these combination systems.

Because the vast majority of all roof drain systems are not connected to the sanitary system, storm drains are typically not trapped. After all, there is no sewer gas to prevent from flowing back into the building. This, in turn, means that the storm drainage system does not require venting, as there is no trap to protect. If you are working on a combination storm/sanitary system, *this is not true*. In a combination system, the storm drain fixtures, such as roof drains and area drains, would have to be trapped, just like any other plumbing fixture.

There are four main types of storm drainage systems. The most common type is the pitched roof with gutters and downspouts. This is usually the most economical system. In the vast majority of instances, the storm water is simply discharged on the ground, where it becomes the responsibility of the civil engineer to direct the water to a suitable point of collection, such as a pond or a public storm sewer inlet. No doubt your home employs exactly this type of system.

The next two types of storm drainage systems involve the installation of flat roofs, which are very common in commercial construction. The more common type uses roof drains at regular intervals to collect the storm water from the roof and transports the water, through pipes, to a suitable point of discharge, such as the ground or a public storm sewer (see Figure 7–1).

Storm Water Systems

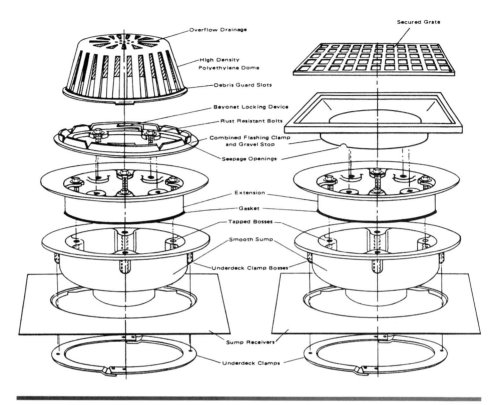

Figure 7–1 Roof drain.
Courtesy Jay R. Smith Mfg. Co.

Another type of storm drainage system used on flat roofs involves the installation of flow control roof drains (see Figure 7—2). These roof drains are very similar to an ordinary roof drain except that the drain contains a device that regulates the flow through the drain and into the piping. In other words, in this type of system, the roof itself acts as a reservoir to hold some portion of the rainwater. The flow control roof drain, over a known period of time, releases a set volume of water into the storm drainage piping system. It goes without saying (but I am going to say it anyway, because of its importance) that this type of roof drain system cannot be designed without the active involvement of the structural engineer. The roof must be designed to carry the added load of the water at the maximum storage level of the roof. For example, 6″ of water on the roof will add approximately 30 pounds per ft^2 (psf) to the roof loading!

The principal advantage to a flow control roof drain system is that it allows you to reduce significantly the size of the roof drain piping system. For

Figure 7—2 Flow control roof drain.
Courtesy Jay R. Smith Mfg. Co.

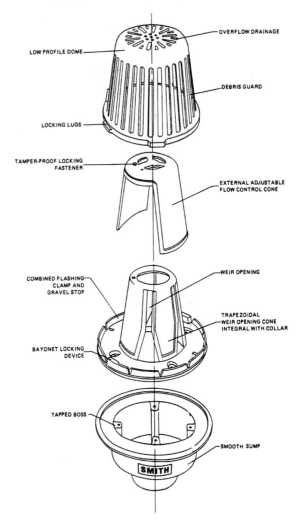

large buildings, this can produce a significant cost savings for the roof drain system, which must be compared to the additional structural costs to determine the overall system economy.

The last type of storm drainage system to be discussed is the site drainage system. This system involves the collection and disposal of storm water from parking lots, driveways, lawns, and other such areas. As stated previously, the flow of surface water across a building site, or a parking lot, is the responsibility of the civil engineer. The goals of adequate site drainage are to keep the surface water from entering the building, eroding the site, or creating a safety or health hazard for people around the site. This is particularly important in those areas of the country where the surface water periodically freezes.

To meet these goals, the civil engineer, the architect, or the owner sometimes deem it necessary to collect the rainwater and convey the rainwater to a city storm sewer, waterway, or retention pond. In these instances, area drains, manholes with grated tops, curb inlets, or other similar structures may be used to collect the rainwater at various points. These structures are then connected, through underground piping, to a point of suitable disposal.

SIZING STORM DRAINAGE SYSTEMS

As with the sanitary and vent systems, the sizing of a roof drain system is a very straightforward procedure, as the following steps indicate:

1. Determine the rainfall rate for the area in which the building is located Usually, this value is set by the local plumbing code. For example, the rainfall rate for Lincoln, NE, is set by the local plumbing code to be 6" per hour. If this rate is not set by code, some research may be required. Often, a local code official will have access to rates deemed appropriate by the local code authority. The U.S. Weather Bureau is also a good source for such data. If appropriate local data are not available, the National Standard Plumbing Code provides a "Maximum Rates of Rainfall" map that may be useful in determining the rainfall rate for your area (see Figure 7–3).

2. Calculate the area of the roof to be drained Do not forget to include the total area that drains into a roof drain. This may include some higher roofs that simply drain rainwater onto a lower roof.

3. Determine the number of roof drains for the system This step must be done in conjunction with the architect. The architect will be able to show you where the low points in the roof structure will be and how many drains that the roof will accommodate. For very small roofs, or even sections of a larger roof, always try to use at least two roof drains. This will ensure that you will still have some drainage, even if one of the roof drains becomes clogged, a fairly common occurrence.

4. Determine the amount of slope available to the storm water piping As with any nonpressurized drainage system, the water is moving by gravity, so it is important that minimum slopes be maintained. Most codes, as well as good plumbing practice, require a minimum slope of at least 1/8" per ft. Naturally, the architect and the owner will expect the storm water piping to be concealed, either above a ceiling or within a wall. Space above a ceiling

Chapter 7

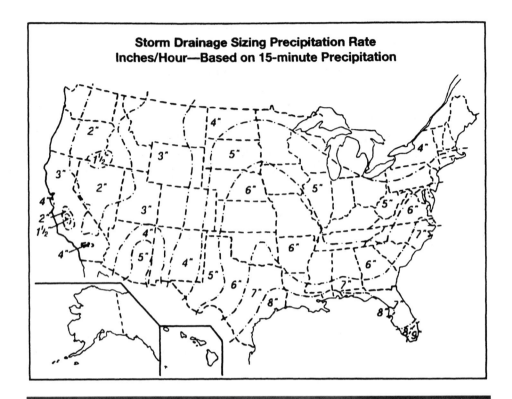

Figure 7-3 Rainfall rate map.
Courtesy *National Standard Plumbing Code,* 1993 Edition. Reprinted with permission from the National Association of Plumbing Heating Cooling Contractors.

is always at a premium (as you will soon experience); therefore, there is always a lot of pressure on the plumbing contractor to "flatten out" the drainage piping by installing the piping at a very slight slope or grade. It has been my experience that exploring an alternative route will solve more problems than simply installing the storm drainage piping at a flat slope, inasmuch as a flat slope generally creates a whole host of maintenance problems due to low flows in the piping. It is well to remember that storm water systems typically carry a large amount of dirt, leaves, and other ground debris which will easily block the flow of water in the piping. Additionally, calculating the slopes available to the storm piping will allow you to determine whether you will be able to discharge the storm water into a city storm main. If the city main is at a higher elevation than the lowest storm main in the building, you will have to either pump the storm water to a higher elevation and then allow it to flow by gravity into the city main, or you will have to discharge the storm water onto the site.

Storm Water Systems

5. Determine the appropriate pipe size using Tables 7–1, 7–2, and 7–3 found in the ANSI-A40 code Table 7–1 provides pipe sizing for vertical storm water pipes, which are sized differently than the horizontal piping which is sized in Table 7–2.

Using values for the area of the roof, the slope of the piping, and the number of roof drains will allow you to size the piping using Tables 7–1 to 7–3. It is very important that you understand that these tables are based on a rainfall rate of 4″ per hour! If you are in a locality that has a higher rainfall rate, you will need to adjust the listed capacity rates. For example, if you live in a town with a rainfall rate of 6″/hr, the capacities listed in the table should be reduced by a factor of 4″/hr divided by 6″/hr, or approximately 0.667. Starting at the furthest point in the system, each section of pipe should be sized based on the area of roof that section of pipe serves. As you travel toward the main,

Table 7–1 Size of vertical conductors and leaders.[a]

Size of Conductor or Leader[b]	Maximum Projected Roof Area	Hydraulic Capacity
in.	ft^2	gpm
2	544	23
2½	987	41
3	1,610	67
4	3,460	144
5	6,280	261
6	10,200	424
8	22,000	913
10	39,600	1,652
12	64,480	2,689

[a]Table 7–1 is based on a minimum rate of rainfall of 4″/hr for a 5-min duration over a 10-yr return period, and on the hydraulic capacities of vertical circular pipes flowing 7/24 full at terminal velocity, computed by the method of NBS Mon 31. Where design rates are more or less than 4″/hr, the figures for drainage area shall be adjusted by multiplying by four and dividing by the local design rate in inches per hour.
[b]The area of rectangular leaders shall be equivalent to that of the circular leader or conductor required. The ratio of width to depth of rectangular leaders shall not exceed 3:1.
Table 25, ANSI A40-1993 Standard, *Safety Requirements for Plumbing.* Reprinted with permission.

Table 7-2 Size of horizontal storm drains and sewers.[a]

Diameter of Drain, in.	Slope					
	⅛ in.		¼ in.		½ in.	
	ft²	gpm	ft²	gpm	ft²	gpm
3	822	34	1,160	48	1,644	68
4	1,880	78	2,650	110	3,760	156
5	3,340	139	4,720	196	6,680	278
6	5,350	222	7,550	314	10,700	445
8	11,500	478	16,300	677	23,000	956
10	20,700	800	29,200	1,214	41,400	1,721
12	33,300	1,384	47,000	1,953	66,000	2,768
15	39,500	2,473	84,000	3,491	119,000	4,946

[a]Table 7-2 is based on a maximum rate of rainfall of 4"/hr for a 5-min duration and a 10-yr return period. Where design maximum rates are more or less than 4"/hr, the figures for drainage area shall be adjusted by multiplying by four and dividing by the local design rate in inches per hour. Gutters other than semicircular may be used provided they have an equivalent cross-sectional area.
Table 26, ANSI A40-1993 Standard, *Safety Requirements for Plumbing*. Reprinted with permission.

the pipe size becomes larger because you are serving a larger section of roof (see Figure 7-4).

In addition to a roof drain system, most jurisdictions also require the installation of an overflow system. An overflow system protects the roof from overloading and collapse, if the primary roof drain system becomes plugged or inoperable. There are two common types of overflow systems: scuppers and overflow drains.

The most common overflow system consists of a "scupper," or outlet, located in the parapet wall, approximately 2" above the inlet of the roof drain system. Naturally, the height of the scupper will be higher for a flow control roof drain system, as the roof will be holding a certain amount of water. The scupper is nothing more than an opening in the parapet wall that will keep the water level on the roof from rising above the bottom of the scupper. Any excess water will simply flow out of the scupper and fall down the side of the building. Clearly, it is important that the scuppers are set at a level that does not exceed the roof loading for the structure. Here again, the structural engineer must be involved in the decision-making process. Additionally, the open area of the scupper must equal the area of the drain openings that it is pro-

Table 7-3 Size of roof gutters.[a]

Diameter of Gutter, in.	Slope							
	1/16 in.		1/8 in.		1/4 in.		1/2 in.	
	ft²	gpm	ft²	gpm	ft²	gpm	ft²	gpm
3	170	7	240	10	340	14	480	20
4	360	15	510	21	720	30	1,020	42
5	625	26	830	37	1,250	52	1,770	74
6	960	40	1,360	57	1,920	80	2,770	115
7	1,380	57	1,950	81	2,760	115	3,900	162
8	1,990	83	2,800	116	3,980	165	5,600	233
10	3,600	150	5,100	212	7,200	299	10,000	416

[a]Table 7–3 is based on a maximum rate of rainfall of 4"/hr for a 5-min duration and 10-yr return period, and on the hydraulic capacity of semicircular gutters. Where design rates are more or less than 4"/hr, the figures for drainage area shall be adjusted by multiplying by four and dividing by the local design rate in inches per hour. Gutters other than semicircular may be used provided they have an equivalent cross-sectional area.
Table 27, ANSI A40-1993 Standard, *Safety Requirements for Plumbing*. Reprinted with permission.

tecting. Making the scupper too small will allow the water level to rise above the level of the scupper, which, in turn, may overload the roof.

The second common type of overflow system involves the installation of a second set of roof drains, usually set approximately 2" above the primary roof drains. There is even a special type of roof drain that has a small standpipe, or dam, built into the drain to provide this 2" of separation (see Figure 7–5).

The piping from the overflow roof drains must be kept separate from the primary roof drains to prevent a plugged storm drain from paralyzing the entire system. The overflow drain piping usually discharges on grade, using the theory that it is only an emergency drain, and, with some luck, will never be used. On rare occasions, the overflow drains are connected to the city storm sewer main. The problem with this, of course, is that if the city main becomes plugged, the entire roof drain system, including the overflow system, is also plugged. It should be readily apparent that this type of secondary overflow system is far more expensive than the scupper system. There are times, however, when a scupper system is simply not feasible or desirable. The overflow pipe system is sized in exactly the same manner as the primary storm piping.

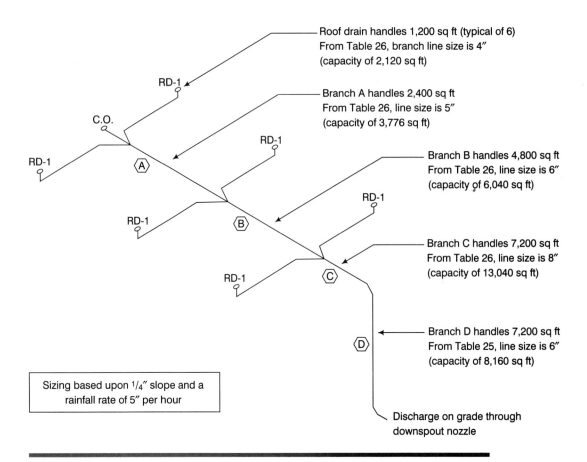

Figure 7–4 Storm system sizing example.

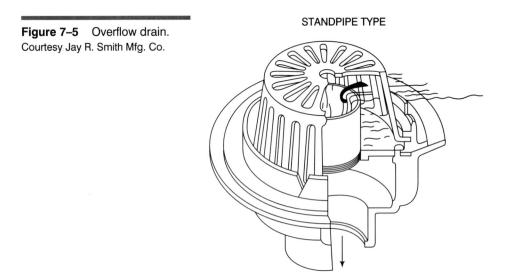

Figure 7–5 Overflow drain.
Courtesy Jay R. Smith Mfg. Co.

The sizing of the site drainage piping is also a very straightforward process although it differs somewhat from the process used to size piping inside the building. The steps are as follows:

1. **Calculate the area (A) to be drained, by surface type** For example, calculate the areas where water is to be collected, but keep the concrete, asphalt, and grassy areas all separate. The areas should be calculated in acres (1 acre = 43,560 ft^2).

2. **Determine the rainfall intensity rate (I)** This is exactly the same rainfall intensity rate that was discussed previously.

3. **Determine the run-off coefficient (C_R)** Use Table 7–4 for some of the more common run-off coefficients.

4. **Determine the amount of slope, or grade, available for the storm water piping.**

Table 7–4 Coefficients of run-off C_R.

Type of Surface	Values of C_R
Impervious surface	0.90–0.95
Steeply sloped barren earth	0.80–0.90
Roofs—flat or nearly flat	0.75–0.95
Asphalt pavement	0.80–0.95
Concrete pavement	0.70–0.90
Rolling barren earth	0.60–0.80
Flat barren earth	0.50–0.70
Rolling meadow	0.40–0.65
Impervious soils, heavy, bare	0.40–0.65*
Deciduous timberland	0.35–0.60
Gravel or rough macadam pavement	0.35–0.70
Impervious soils with turf	0.30–0.55*
Conifer timberland	0.25–0.50
Orchard	0.15–0.40
Gravel surfaces	0.15–0.30
Slightly pervious soils	0.15–0.40*
Rolling farmland	0.15–0.40
Flat farmland	0.10–0.30
Slightly pervious soils with turf	0.10–0.30*
Parks, cultivated land, lawns, etc., depending on slope	0.05–0.30
Moderately pervious soils	0.00–0.20*
Wooded areas	0.01–0.20

Table 16–4, *Concrete Pipe Handbook,* 1959, p. 281. Reprinted with permission of the American Concrete Pipe Association.

5. **Using the formula,**

   Flow (in cubic ft/sec) = $C_R \times I \times A$,

calculate the flow of storm water from the area. This calculation will be necessary for each area that has a differing run-off coefficient.

6. **Determine the pipe size using a Manning Chart** The pipe size is determined by plotting the flow rate line and the slope line onto the Manning chart. The pipe size is read from the intersection of these two lines. For example, for a 20,000-ft² concrete parking lot, assuming a rainfall intensity rate of 4″/hr and a maximum pipe slope of ⅛″ per foot, the calculations would be as follows:

   A = 20,000 ft² / 43,560 ft² per acre = 0.46 acre

   I = 4″/hr of rainfall

   C_R = 0.80 (average between 0.70 and 0.90)

Therefore,

   Flow = 0.46 acre × 4″/hr × 0.80 = 1.47 ft³/sec (CFS)

Plotting the vertical slope line (1/8″/ft equals 0.0104′ of fall per foot, which, in turn, equals 1.04′ of fall per 100 feet) and the horizontal flow line (1.47 CFS) on the Manning chart will produce an intersection which is between the 8″ pipe diameter and the 10″ pipe diameter. Usually, most engineers would select the larger of the two diameters, which, in this case, is 10″. By extending the horizontal flow line to intersect the 10″ pipe diameter line, we can read the velocity of the water in the pipe to be approximately 2.5 ft/sec, which is adequate for this application (see Figure 7–6).

As expected, there is a list of rules for storm water piping on the site.

1. **The minimum size of storm water mains should be 12″** The reason for the rather large minimum sizes for piping located on the site is that the site storm water system will naturally pick up a lot of debris, dirt, and other items in addition to the storm water. These minimum sizes will keep your site storm water system from constantly becoming plugged with this debris.

2. **The minimum size of storm water branches should be 8″.**

3. **The minimum velocity in a storm water pipe should be 2.5 ft/sec.**

4. **The minimum earth cover above a storm water line should be 2.5′** The minimum cover for a pipe located below a roadway or driveway is a function of material selected. Be aware that some materials have both a minimum and a maximum value for the cover over a pipe.

5. **Provide a manhole or an inlet box at each change in direction or at each change in slope.**

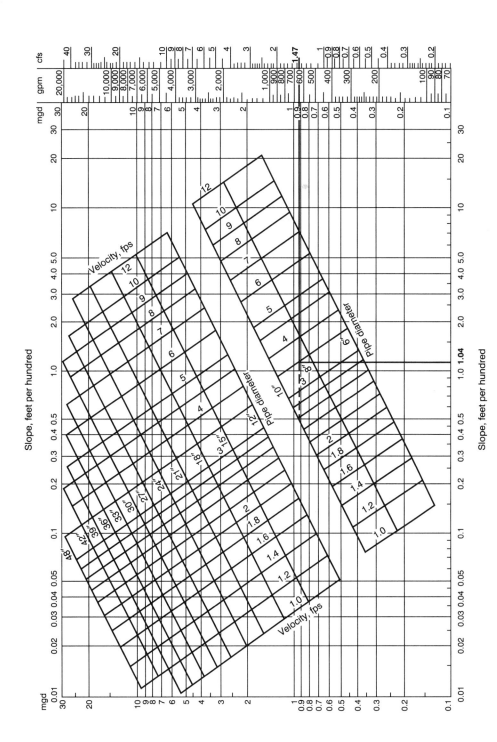

Figure 7-6 Manning's pipe flow chart.
N - 0.013 for 12" pipe and larger; N - 0.014 for 10" pipe and smaller.
From Technical Manual TM 5-814-1, *Sanitary Engineering, Sanitary and Industrial Waste Sewers*, August 1966, Department of the Army, United States of America.

REVIEW QUESTIONS

1. What purposes does a storm system serve for a building?
2. Why should you install at least two roof drains in a roof or roof section?
3. What is the purpose of an overflow system? What types of overflow systems exist?
4. What special consideration is important when installing a flow control roof drain system?
5. What area, in square feet, can a 6" horizontal storm drain pipe handle, if it is installed at a ¼" per ft slope, in a city with a rainfall rate of 5"/hr?
6. Why are combination sanitary/storm systems prohibited in most cities?
7. What size of semicircular roof gutter is appropriate for a rectangular roof 30' × 65' in area, installed in a city with a rainfall rate of 6"/hr, assuming a slope of ⅛" per ft? How does the answer change if the slope is increased to ¼" per ft?
8. Size the roof drains and roof drain piping for the following rectangular flat roof, assuming a rainfall rate of 5"/hr and a slope of ⅛" per ft.

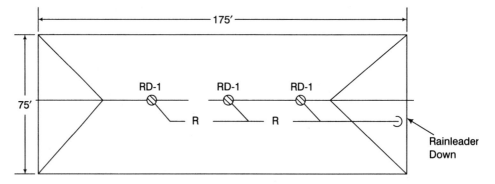

9. Assume a 300' × 400' asphalt parking lot has two storm water inlets. If the parking lot is located in a city with a rainfall rate of 6"/hr and the piping has a slope of ¼" per ft, what size of pipe must serve each storm water inlet?

CHAPTER 8

Domestic Water System

The purpose of the domestic water system is to provide a safe, healthy supply of potable water, at an adequate temperature and pressure, to all plumbing fixtures within a building. Fundamentally, the water distribution system differs materially from the drain, waste, and vent system in a variety of ways, which this chapter discusses.

PRESSURE DIFFERENTIAL

Water distribution systems flow due to a pressure differential, not by gravity. Said another way, water flows in a pipe from a point of high pressure to a point of lower pressure. Pressure affects the distribution of the domestic water system in several ways. Firstly, if the water pressure is not adequate at the plumbing fixture, the plumbing fixture will not work properly. We have all probably experienced a shower or lavatory where the water just barely trickled out of the fixture. Needless to say, this is very undesirable.

Secondly, most plumbing fixtures are designed to accept water at a pressure not to exceed 80 psi. In fact, most codes prohibit the water distribution system from exceeding 80 psi, and for good reason. I have seen plumbing fixtures literally blown off the wall when accidentally supplied with water of 100 psi or more. This situation is even more undesirable than inadequate water pressure.

Thirdly, water flowing through a piping system loses pressure due to friction losses and differences in elevation. The pressure is not lost, of course; it is merely transferred into another form of energy. It is lost only in that we cannot use the pressure to service the fixture.

The factors that affect these pressure losses, in a domestic water system, are as follows:

1. **The starting pressure** Usually, this is the pressure at the city main, located in front of the building under the street. However, it may also be the pressure at the outlet of a booster pump or the outlet of an elevated pressure tank.

2. **The ending pressure** This is the minimum pressure required at the plumbing fixture to make the fixture work properly. Common minimum pressures are

   A. Tank-type water closets   15 psi
   B. Flushometers   25 psi
   C. Lavatories   8 psi

3. **The pressure necessary to lift the water to a new elevation** A bathroom on the third floor of a building, for example, takes some energy to lift the water to that elevation. Remember that 1 psi of pressure lifts water 2.307'.

4. **The friction losses through pipe, fittings, and valves** As you will see in the next section, we always assume a fairly rough pipe to allow for the resistance due to internal pipe or surface roughness (and the resulting loss of pressure) that always occurs with flow and steadily increases over time.

At this point, you are probably thinking, "Well, at least I don't have to be concerned about slopes and grades with domestic water piping!" As construction managers, you should be so lucky. Slopes and grades do not enter into the design equation, as they do in a DWV system; however, it is always a good idea to slope all water piping to a low point, on one end of the system or the other, to facilitate draining down the system to repair or modify the water distribution system. It also greatly assists in bleeding air out of the system to prevent "air locking," which, in turn, will obstruct the flow of water.

From this discussion, we can conclude that pressure loss in the water distribution system is one of the main design criteria used in sizing the water piping.

NOISE

Noise is a constant problem in designing a proper water distribution system. Of all the complaints you will hear about plumbing systems, pressure fluctuations and noise constitute the largest percentage of the problems you will encounter.

Noise problems manifest themselves in two ways. First, we have to deal with the problem of "water hammer." Water hammer is a shock wave inside the water distribution piping that is created when a valve quickly shuts off the flow of water (see Figure 8–1).

The shock wave creates a pressure rise that is approximately 60 times the velocity of the water in the pipe. Considering that most engineers design a water system to maintain a flow velocity of 5 to 10 ft/sec (fps), it follows that this shock wave can easily reach a pressure of 600 psi. Further, this shock wave is traveling at approximately 4,000 to 4,500 fps. It is easy to see that a shock wave can create a tremendous amount of noise, pipe movement, and damage in a piping system.

The problem of water hammer often occurs in plumbing systems because most plumbing fixtures, except tank-type water closets, use a quick-closing valve. For example, think about how quickly you turn a kitchen faucet on and off. Although this doesn't seem like a big deal on the surface, it takes very little to produce a significant shock wave in the water distribution system. Because it is impossible to eliminate quick-closing valves in a plumbing system, we use two different devices to "dampen" the shock wave as it is created. These two devices are air chambers and shock absorbers. Both allow the shock wave to be dissipated through the compression of a gas that absorbs the energy of a shock wave (see Figures 8–2 and 8–3).

Both the air chamber and the shock stop have their advantages and disadvantages. Air chambers tend to become "water logged" and thus lose their air cushion, which, of course, eliminates their effectiveness at dissipating a shock wave. Shock stops, on the other hand, have a tendency to become clogged with calcium carbonate in those localities with hard water. The calcium carbonate prohibits the bellows from expanding and absorbing the shock wave. Additionally, I have recently read a few reports where some hospitals have found legionnella bacterium contamination in the shock stops. Both air chambers and shock stops need to be carefully designed and installed if they are to prevent water hammer damage.

It should also be noted that the velocity of the water in the piping can cause a considerable amount of noise, particularly in thinner wall copper pipe, such as Type M. Most design engineers, and even most plumbing codes, limit the velocity of the water in the piping to 10 fps. As stated before, most engineers also endeavor to maintain a water velocity of at least 5 fps in order to keep pipe sizes as small as possible and, therefore, economical. The important point to

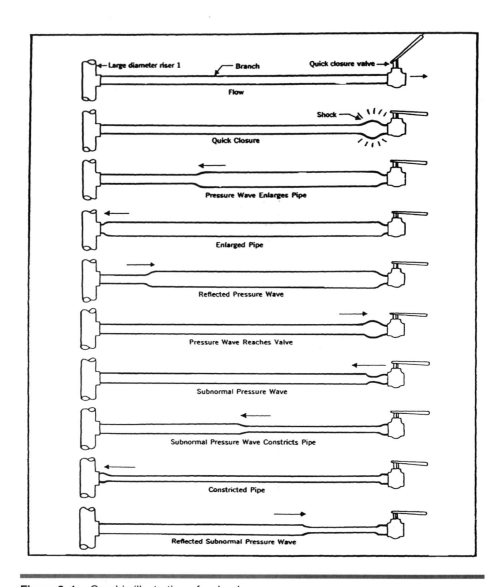

Figure 8–1 Graphic illustration of a shock wave.
July 1992, Plumbing and Drainage Institute.

Figure 8–2 Shock absorber. Reprinted courtesy of Josam Manufacturing Company.

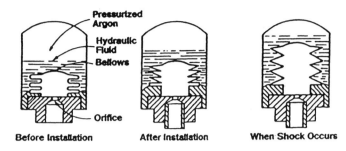

Before Installation   After Installation   When Shock Occurs

Illustration courtesy of Josam Mfg. Co.

Before installation, engineered shock absorber bellows are held in fully compressed position by pressurized gas in upper part of chamber. After installation line pressure extends and flexes bellows until pressures inside and outside are equalized. When hydrostatic shock occurs, increased line pressure extends bellows, absorbing the shock. After shock, bellows returns to normal installed position.

Figure 8–3 Air chambers. July 1992, Plumbing and Drainage Institute.

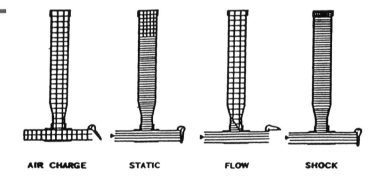

AIR CHARGE   STATIC   FLOW   SHOCK

remember is that velocity and pressure drop are related. The higher the velocity, the higher the pressure drop through the pipe, fittings, and valves.

From this discussion, we can conclude that the velocity of the water in the piping system is another major criterion to consider in the design of the water distribution system.

TEMPERATURE

The temperature of the water is the third major criterion to consider in the design of the water distribution system. The temperature of the water required varies greatly with the needs and uses within the system. For example, most health codes require 180°F hot water supplied to a commercial dishwasher; yet, at the same time, the hand-washing sinks in the same commercial establishment require hot water at only 105°F (see Figure 8–4).

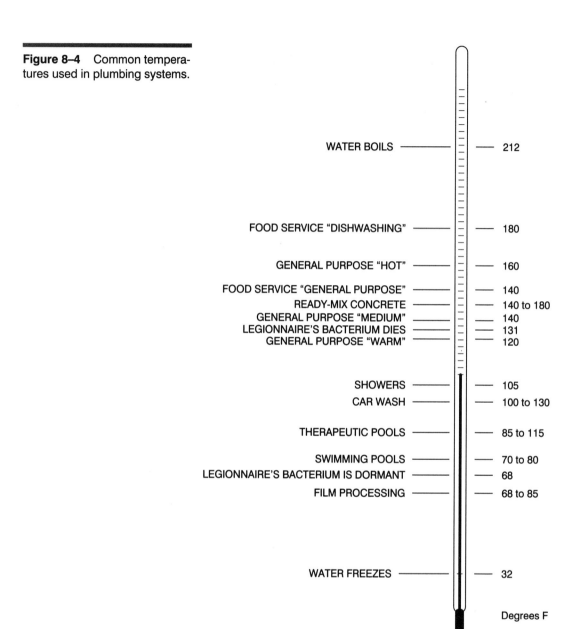

Figure 8-4 Common temperatures used in plumbing systems.

Following are four special considerations to keep in mind when dealing with domestic hot water:

1. Hot water burns people   In our litigious society, you can count on being one of the parties sued if you participated in the construction of a building where someone was burned by the domestic hot water system. Domestic hot water, as Table 8–1 shows, can create this type of tragedy very quickly indeed.

2. Hot water is corrosive   For every 20 degrees that you raise the water temperature in the tank of a hot water heater, you double the corrosive qualities of the water. This effect is even more pronounced if the water is already somewhat corrosive.

3. Legionnella bacterium (Legionnaire's Disease) can breed in a domestic hot water system   Legionnella bacterium becomes dormant at 68°F and dies at 131°F. Its point of highest reproduction is around 110° to 120°F, which, unfortunately, is the temperature setting of most hot water heaters.

4. Electrolysis develops between dissimilar metals   The process of electrolysis (electron transfer from one metal to another metal) is highly temperature dependent. That is, as the temperature increases, the electron transfer from one metal pipe to another dissimilar metal pipe is increased. As the electron transfer increases, the metal pipe losing the electrons corrodes.

To achieve temperature stability, some hot water systems require the installation of a hot water recirculation system. A hot water recalculation system does nothing more than pump a small percentage of the total domestic hot water demand to the farthest point in the domestic hot water system. The recirculation system ensures that hot water is available at every plumbing fixture "on demand," or as you open the faucet. Without a recirculation system, the hot water in the distribution piping has a tendency to cool off during peri-

Table 8–1   Temperature/time/burn.
Courtesy *Plumbing Engineer* magazine.

Temperature of Water in Degrees F	Minimum Time for First-Degree Burn	Minimum Time for Second- or Third-Degree Burn
111.2	5 hours	7 hours
116.6	35 minutes	45 minutes
118.4	10 minutes	14 minutes
122.0	1 minute	5 minutes
131.0	5 seconds	25 seconds
140.0	2 seconds	5 seconds
149.0	1 second	2 seconds
158.0	—	1 second

ods of low use, so that when you wanted to use a lavatory or a sink, you would be forced to run lots of tepid water down the drain before you got any really warm water. This type of temperature problem manifests itself only in hot water distribution systems of significant length. Accordingly, most codes only require hot water recirculation systems in buildings where the length of the hot water distribution system exceeds 100'.

PUBLIC HEALTH

Hot water and healthy water go hand in hand. There has been a lot of controversy lately about what really constitutes healthy water. Every water system, regardless of its source, contains a number of different chemicals, in addition to H_2O. A typical water analysis would probably alarm most individuals not familiar with higher chemistry and, in fact, is undoubtedly one of the reasons for the recent growth in bottled water sales (see Figure 8–5).

Another health-related topic that is frequently in the news concerns lead in the drinking water. Historically, the plumbing industry has used lead in many different ways: in lead-based solders for making copper piping joints, in lead piping for water services to homes and businesses, and in brass piping and faucets in fixtures. Recent regulations prohibit the use of lead-based solders and lead water pipe; however, the industry is still searching for an acceptable, economical replacement for brass faucets and piping.

Just as we do not want the city-supplied water to contaminate the water in our homes, we also do not want the "used" water from our homes to contaminate the city main. As a result, most city codes require the building owner to protect the city main from contamination. This is accomplished in two ways.

In a commercial application, most city codes require the installation of a backflow preventer on the water service to the building, immediately after the service enters the building. Typically, the backflow preventer is installed immediately downstream of the water meter. The purpose of the backflow preventer is to prevent domestic water in the building from flowing backwards and returning to the city main. It needs to be stressed that a backflow preventer will protect the city main from contamination; however, *it does not protect the occupants of the building from contamination within the building!* All too often I have had a well-intentioned maintenance manager explain to me that contamination is not a problem because there is a backflow preventer in the basement. There are any number of possible backflow conditions that might contaminate the water distribution system within a building, including lawn sprinkler connections, threaded service sink faucets, boilers,

Figure 8–5 Water analysis.

	Filtered Water Ashland Plant
pH	7.5*
Total Alkalinity (CaCO$_3$)	152 mg/l
Total Hardness (CaCO$_3$)	180 mg/l = 10.50 grains
Total Dissolved Solids	286 mg/l
Sodium (Na)	30 mg/l
Calcium (Ca)	58 mg/l
Manganese (Mg)	<0.05 mg/l
Iron (Fe)	<0.1 mg/l
Fluoride (F)	0.74 mg/l
Chloride (Cl)	16 mg/l
Sulfate (SO$_4$)	79 mg/l
Nitrate (N)	0.3 mg/l
Magnesium	7 mg/l*

EXPLANATION OF RESULTS OF CHEMICAL WATER ANALYSIS

Nitrate in excess of 10 mg/l (as nitrate-nitrogen) is of health significance to infants under six months of age. High nitrate water should not be used for infant formulas or in infant food. Higher nitrate content apparently is tolerated by all other ages, but nitrate-nitrogen of over 50 mg/l is very undesirable.

Fluoride is important in development of teeth in children. The optimum fluoride content to assist in control of tooth decay is 0.9 to 1.5 mg/l. Concentrations over 3.0 mg/l may cause darkening of the tooth enamel and possible other undesirable effects.

pH is a measure of the acidity (below pH 7) or alkalinity (above pH 7) of water. Absolutely pure water has a pH of 7, but this is an ideal value and is rare. Normally, in Nebraska, well water pH will be between 6.5 and 8.0.

and dozens of other similar applications. Therefore, you have to be very alert to protecting the water distribution piping within the building from backflow and back-siphonage. Figure 8–6 lists a number of backflow prevention devices and their applications.

Another means of protecting the domestic water distribution system is by providing an air gap between the water outlet and the flood level rim of the fixture. This air gap ensures that water will not rise in the fixture into the water supply outlet. Table 8–2 lists the prescribed air gap distances for a variety of plumbing fixtures.

Air gaps are very effective at preventing this type of contamination and are usually the first line of defense, as far as the occupants are concerned.

Figure 8–5 *continued*

Hardness is a measure of the soap-consuming capacity of water; that is, the more soap required to produce a lather, the harder the water. Hard water causes rings on bathtubs and sinks, chemical deposits in pipes, and poor laundry results. Hard water, however, is not considered to be detrimental to health. The hardness of water may be removed by a zeolite (exchange resin) water softener. Zeolite softened water may be of concern to persons on a low salt diet for medical reasons. Hardness is reported in mg/l as calcium carbonate. To convert to grains per gallon divide the mg/l value by 17.1. A survey of hardness of Nebraska water supplies showed:

less than 100 mg/l	4%	soft water
100 – 200 mg/l	16%	reasonably soft
200 – 300 mg/l	40%	average hardness
300 – 400 mg/l	23%	very hard
over 400 mg/l	17%	extremely hard

Calcium and Magnesium contribute to the majority of the hardness of water, separate values are of minor concern.

Sodium may be of health significance to persons on a low salt diet for medical reasons. Sodium is increased in water that has been passed through a zeolite softener. No established limit is set for sodium content, but more than 100 mg/l is undesirable.

Iron and Manganese are nuisance chemicals which cause stains on clothes and plumbing fixtures. Iron causes reddish-brown stains while manganese deposits are gray. Iron in excess of 0.3 mg/l and manganese in excess of 0.2 mg/l can cause problems.

Sulfate content in excess of 300 to 500 mg/l may give a bitter taste to water and have a laxative effect on persons not adapted to the water.

Total Solids is the total amount of material remaining after the evaporation of the water. Values of 500 mg/l are satisfactory and up to 1000 mg/l can be tolerated with little effect.

All results in milligrams per liter (with the exception of pH).
Courtesy City of Lincoln, NE.

Figure 8-6 Backflow prevention device application chart.

Device Type & Purpose	Description	Installed at	Examples of Installations	A.S.S.E. Applicable Standards
Reduced Pressure Principal Backflow Preventer For health hazard cross connections	Two independent check valves with intermediate relief valve. Supplied with gate valves, test cocks, inlet and outlet unions, and strainer	All cross connections subject to backpressure and back-siphonage where there is a high potential health hazard from contamination. Continuous pressure	Main supply lines Commercial boilers Hospital equipment Processing tanks Laboratory equipment Air conditioning Waste digesters Lawn sprinklers Sewerage treatment Cooling towers	A.S.S.E. No. 1013 Sizes ¾"–10"
Double Check Valve Assembly For low hazard cross connections	Two independent check valves. Supplied with gate valves, test cocks, inlet and outlet unions, and optional strainer	All cross connections subject to backpressure where there is a low potential hazard or nuisance. Continuous pressure	Main supply lines Food cookers Tanks and vats Lawn sprinklers Fire lines Commercial pools	A.S.S.E. No. 1015 Sizes ½"–10"
Dual Check Valve Assembly For low hazard cross connections	Two independent check valves, supplied with inlet union and optional strainer	Cross connections subject to back pressure where there is a low potential hazard or nuisance. Continuous pressure	Residential water services Individual outlets	A.S.S.E. No. 1024 Sizes ½"–1"

Figure 8-6 continued

Device Type & Purpose	Description	Installed at	Examples of Installations	A.S.S.E. Applicable Standards
Backflow Preventer with Intermediate Atmospheric Vent — For low hazard cross connections in small pipe sizes	Two independent check valves with intermediate vacuum breaker and relief valve. Supplied with inlet and outlet unions and integral strainer	Cross connections subject to backpressure and back siphonage where there is a low hazard. Continuous pressure	Residential boilers Cooling towers Dairy equipment Photo laboratory equipment Residential	A.S.S.E. No. 1012 Sizes ½"–¾"
	Special model for carbonated beverage vending machines	To prevent backflow of carbon dioxide gas and carbonated water into the water supply system to vending machines	Carbonated vending machine	Special Approvals
Atmospheric Vacuum Breakers — For low and high hazard cross connections	Single float and disc with large atmospheric port	Cross connections not subject to backpressure or continuous pressure. Install at least 6" above fixture rim. Provides protection against back siphonage	Lawn sprinklers Process tanks Dishwashers Soap dispensers Washing machines	A.S.S.E. No. 1001 ANSI. A112.1.1 Sizes ¼"–4"
Anti-Siphon Pressure Type Vacuum Breakers — For low and high hazard cross connections	Spring loaded single float and disc with independent first check. Supplied with test cocks and gate valves	This valve is designed for installation in a continuous pressure potable water supply system 12" above the overflow level of the container being supplied, with no back pressure. Provides protection against back siphonage	Commercial laundry machines Swimming pools Chemical plating tanks Photo tanks Large toilet and urinal facilities Heat exchangers Degreasers Live stock water systems	A.S.S.E. No. 1020 ANSI A112.1.7 Sizes ½"–10"
Hose Connection Vacuum Breakers — For residential and industrial host supply outlets	Single check with atmospheric vacuum breaker vent	Install directly on hose bibbs, service sinks, and wall hydrants. Not for continuous pressure	Hose bibbs Service sinks Hydrants	A.S.S.E. No. 1011 ANSI A112.1.3 Size ¾" ANSI ASSE 1019

Courtesy Plumbing Engineer magazine.

Table 8–2 Minimum air gaps for plumbing fixtures.

Fixtures	Minimum Air Gap	
	When not affected by near wall[a], in.	When affected by near wall[b], in.
Lavatories and other fixtures with effective openings not greater than ½ in. diameter	1	1½
Sink, laundry trays, gooseneck bath faucets, and other fixtures with effective openings not greater than ¾ in. diameter	1½	2¼
Over rim bath fillers and other fixtures with effective openings not greater than 1 in. diameter	2	3
Drinking water fountains—single orifice not greater than 7⁄16 in. in diameter; multiple orifices having total area of 0.150[1] in. (area of circle: 7⁄16 in. diameter)	1	1½
Effective openings greater than 1 in. of effective opening	2 times diameter of effective opening	3 times diameter

Table 6, ANSI A40-1993 Standard, *Safety Requirements for Plumbing*. Reprinted with permission.

EXPANSION AND CONTRACTION

Expansion and contraction of the domestic water piping is also a consideration in the design of the water distribution system. Both copper pipe and plastic pipe have a fairly high coefficient of expansion. This means that the piping will "grow" and "shrink" in length as the temperature of the water increases or decreases. You may be inclined to think that once the domestic hot water system is activated, the water will always be at one constant temperature, and there will be no further expansion or contraction of the piping. This is not the case. As we discussed previously, at times of no flow or even low flow, the domestic hot water will simply sit in the piping and give up its heat to the surrounding environment, which, of course, lowers the water temperature. Although domestic hot water lines are typically insulated with fiberglass insulation to retard this heat loss, the heat loss from the piping can still be significant under the right conditions. Additionally, the hot water heater in any system will periodically

106 Chapter 8

have to be taken out of service for maintenance and repair. When this occurs, the domestic hot water system quickly returns to the ambient temperature.

In most applications, expansion and contraction are accounted for with the installation of expansion joints or expansion loops. Both of these devices allow the system to grow or shrink without placing undue stress on the pipe or fittings (see Figure 8–7).

Figure 8–7 (a) One type of drainage line expansion joint for vertical use; (b) expansion loop for water line—plan view.

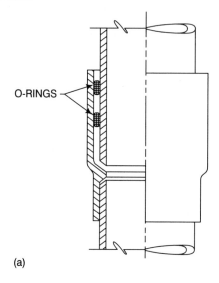

(a)

NOTE: THE METHOD OF PROVIDING THE APPROPRIATE STRESS RELIEF IS BASED UPON THE TYPE OF MATERIAL USED, THE TEMPERATURE CHANGE IN THE PIPING, AND THE NUMBER OF DIRECTION CHANGES MADE IN THE PIPING.

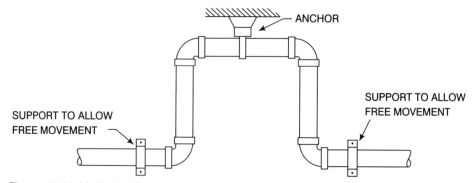

Figure 4.2.16(a,b), *National Standard Plumbing Code,* 1993 Edition. Reprinted with permission from the National Association of Plumbing Heating Cooling Contractors.

SIZING THE DOMESTIC WATER SYSTEM

Now that we have a better understanding of the fundamentals of the domestic water distribution system, we can begin to size the water piping in a simple system. Assume that the plumbing fixtures have already been selected and located during the design of the drain, waste, and vent system.

The steps necessary to size the domestic water piping are as follows:

1. Draw an isometric diagram of the bathroom or system to be sized This is exactly the same type of isometric diagram discussed in the design of the DWV system. The only difference is that the water piping is usually depicted with the following symbols:

Cold water —— - ——
Hot water —— - - ——
Hot water circulation —— - - - ——

2. Assign water supply fixture units (WSFU) to each plumbing fixture The WSFU rating for each fixture was also discussed previously, and the values for the more common plumbing fixtures are given in Table 2–2, in Chapter 2. It is very important to note that there is a difference between a "public" fixture and a "private" fixture. Remember, the WSFU rating is only a probability, and a public fixture clearly has a higher probability of use than a private fixture. Designating the fixtures as "public" or "private" is an issue that must be resolved at the beginning of the design. Another issue to cover is the difference between a system that is primarily comprised of flush tanks versus a system that is primarily comprised of flushometers. As we discussed previously, a flushometer uses a great deal more water than a flush tank. For this reason, the table and chart used to convert WSFU's to gallons per minute (gpm) are always divided between flush tank systems and flushometer systems. Because a flushometer is only used on cold water systems (only cold water is used to flush a water closet), the hot water system is always sized based on the flush tank curve or tables. If the branch or section of piping that you are sizing contains a flushometer, that section or branch would be sized using the flushometer table or curve.

3. Starting at the most remote point from the water meter, and traveling toward the water service entrance, add all of the WSFU values It is imperative to remember that WSFU values are additive, because they represent probabilities. As probabilities, they have diversity (the probability that not all of the fixtures will be used at the same time) contained within the value. Said another way, the higher the WSFU value, the higher the probability that not all of the fixtures will be in simultaneous use. This makes sense if you consider a bank of 10 water closets in a large, public bathroom and ask yourself, "What are the odds that all 10 water closets will be used at exactly the same time?" Obviously, the odds are low indeed. The flow values, such as gallons per minute (gpm) are not additive, as they are not probabilities. If you do try to add gpm flows, you will obtain an artificially high answer. The exception to this rule would be a continuous flow, such as a processing tank that uses a constant, steady flow of 5 gpm. In this instance, the continuous flow is added to the flow converted from WSFU value, as described in Step 4.

   4. **Convert the WSFU values to gpm values using either a Hunter Curve or a conversion table (see Figure 8–8 and Table 8–3).**
   Note that the Hunter Curve has two curves: one for flush tanks and one for flushometers.

   5. **Select the pipe sizing chart for this application** In the vast majority of cases, you will select a chart for copper piping, as most water distribution systems are designed and specified for copper. Always use a sizing chart for the material selected for the water distribution system. Figures 8–9 through 8–17 represent most of the major material types currently used for domestic water piping. All of these charts are based upon fairly rough piping, which provides some factor of safety in determining the pipe size. This is only appropriate as the inside of the pipe will, in fact, become more rough as it ages, accumulating a coating of precipitates from the water.

   6. **Select the appropriate pressure drop and water velocity for the piping** As we discussed earlier, you must determine the total amount of pressure drop available for the entire distribution system. The total amount of pressure drop and the velocity of water in the piping are related. The higher the velocity of the water, the greater the losses that are created by friction and, therefore, the higher the pressure drop in the system. That is why these two decisions have been combined into one step. Also, remember that most engineers try to keep the water velocity between 5 ft/sec and 10 ft/sec. With these parameters in mind, I have drawn a line on the water sizing graph for copper pipe (Figures 8–9 and 8–10) that maintains a flow velocity between 10 fps and 4 fps while varying the pressure loss from 2 psi per 100' of pipe to 7 psi per 100' of pipe. This target line will produce a good economical average for most simple pipe-sizing situations.

   7. **Cross-check individual supply line sizes against code minimum** The minimum supply sizes for most common plumbing fixtures is given in Table 2–2, in Chapter 2.

   8. **Cross-check available pressure at the highest and/or most distant plumbing fixture to ensure that the pressure is sufficient** Depending on the outcome of this cross-check, you may have to increase the pipe sizing to lower the pressure drop in the system and, accordingly, increase the pressure available at the last fixture.

If it is simply not possible to provide enough pressure at the highest or most distant fixture, you will have to design and install a pressure booster system.

Figure 8–17 illustrates laying out and sizing a simple bathroom using these principles.

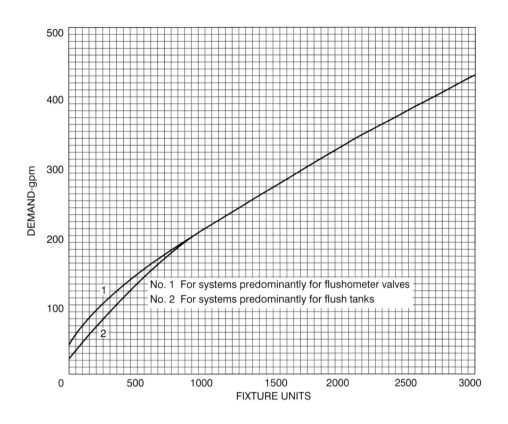

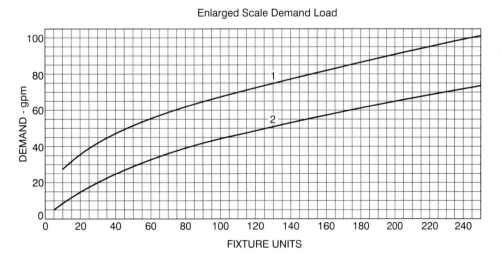

Figure 8–8 Hunter Curve.
Figure 2, ANSI A40-1993 Standard, *Safety Requirements for Plumbing*. Reprinted with permission.

Table 8–3 Estimating demand.

Supply Systems Predominantly for Flush Tanks Load		Supply Systems Predominantly for Flushometer Valves Load	
Water Supply Fixture Units	Demand, gpm	Water Supply Fixture Units	Demand, gpm
6	5		
8	6.5		
10	8	10	27
12	9.2	12	28.6
14	10.4	14	30.2
16	11.6	16	31.8
18	12.8	18	33.4
20	14	20	35
25	17	25	38
30	20	30	41
35	22.5	35	43.8
40	24.8	40	46.5
45	27	45	49
50	29	50	51.5
60	32	60	55
70	35	70	58.5
80	38	80	62
90	41	90	64.8
100	43.5	100	67.5
120	48	120	72.5
140	52.5	140	77.5
160	57	160	82.5
180	61	180	87

Supply Systems Predominantly for Flush Tanks Load		Supply Systems Predominantly for Flushometer Valves Load	
Water Supply Fixture Units	Demand, gpm	Water Supply Fixture Units	Demand, gpm
200	65	200	91.5
225	70	225	97
250	75	250	101
275	80	275	105.5
300	85	300	110
400	105	400	126
500	125	500	142
750	170	750	178
1,000	208	1,000	208
1,250	240	1,250	240
1,500	267	1,500	267
1,750	294	1,750	294
2,000	321	2,000	321
2,250	348	2,250	348
2,500	375	2,500	375
2,750	402	2,750	402
3,000	432	3,000	432
4,000	525	4,000	525
5,000	593	5,000	593
6,000	643	6,000	643
7,000	685	7,000	685
8,000	718	8,000	718
9,000	745	9,000	745
10,000	769	10,000	769

Table 8, ANSI A40-1993 Standard, *Safety Requirements for Plumbing*. Reprinted with permission.

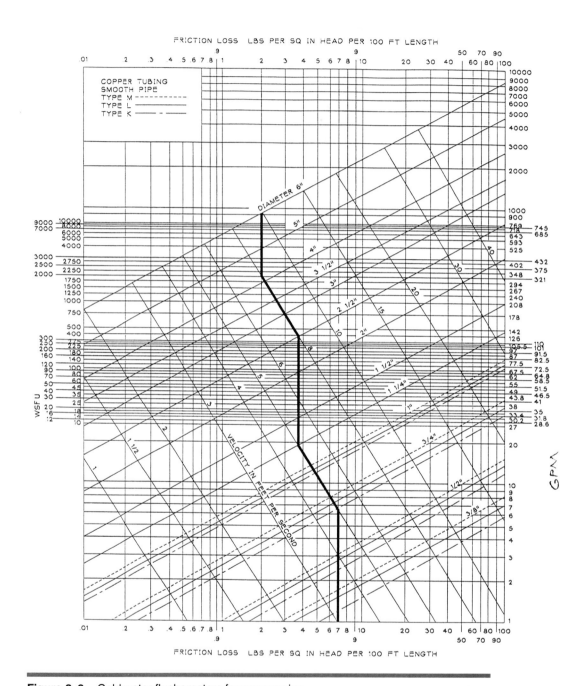

Figure 8–9 Cold water flushometers for copper pipe.

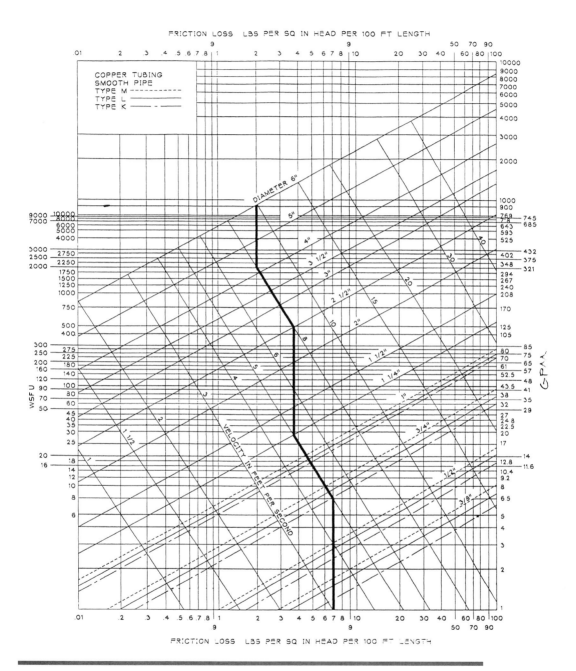

Figure 8–10 Hot water and/or cold water without flushometers for copper pipe.

113

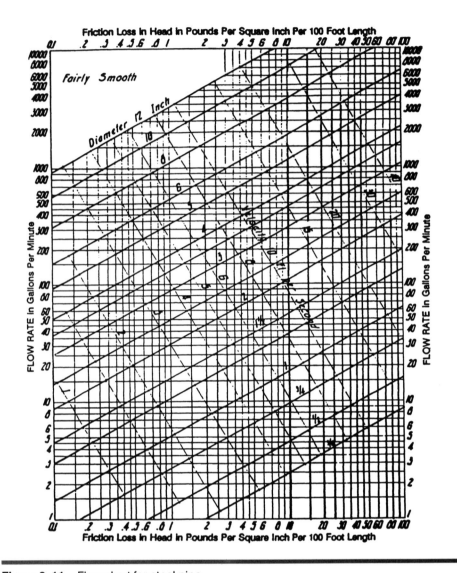

Figure 8–11 Flow chart for steel pipe.
Figure 4, ANSI A40-1993 Standard, *Safety Requirements for Plumbing*. Reprinted with permission.

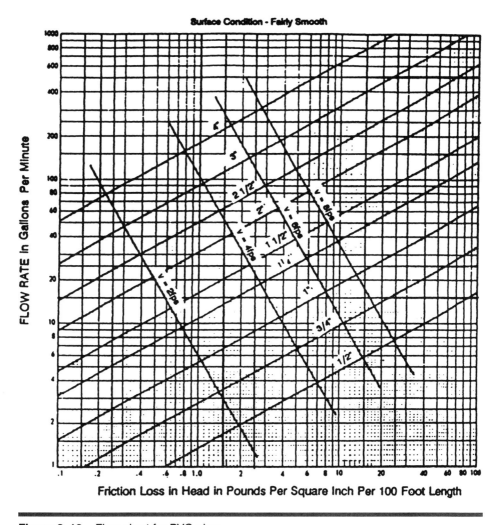

Figure 8–12 Flow chart for PVC pipe.
Figure 5, ANSI A40-1993 Standard, *Safety Requirements for Plumbing*. Reprinted with permission.

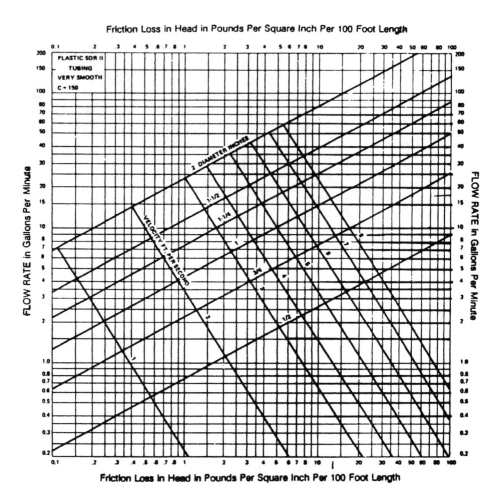

Figure 8–13 Flow chart for CPVC tubing.
Figure 6, ANSI A40-1993 Standard, *Safety Requirements for Plumbing*. Reprinted with permission.

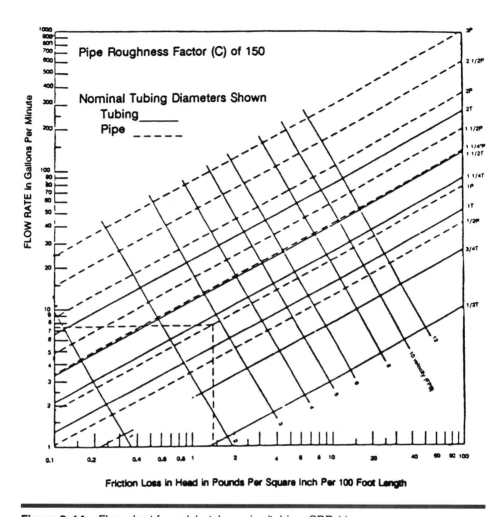

Figure 8–14 Flow chart for polybutylene pipe/tubing, SDR 11.
Figure 7, ANSI A40-1993 Standard, *Safety Requirements for Plumbing*. Reprinted with permission.

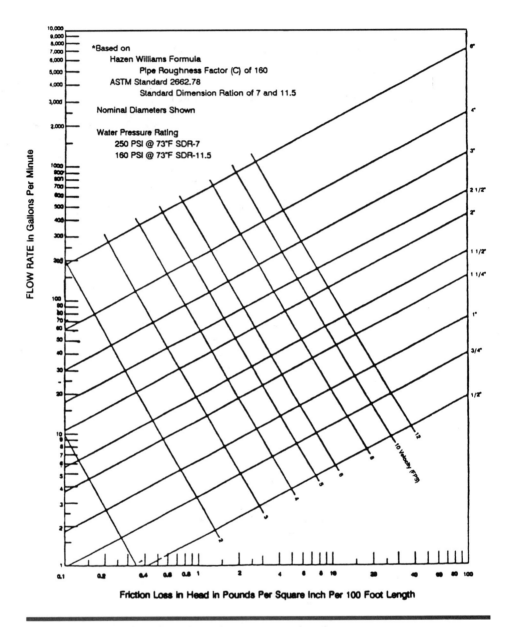

Figure 8–15 Flow chart for polybutylene water service pipe.
Figure 8, ANSI A40-1993 Standard, *Safety Requirements for Plumbing*. Reprinted with permission.

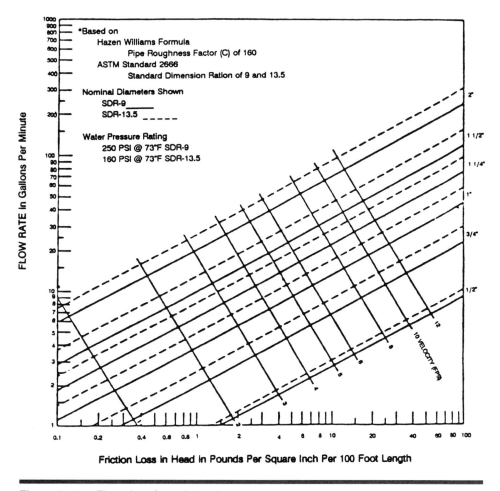

Figure 8–16 Flow chart for polybutylene water service tubing.
Figure 9, ANSI A40-1993 Standard, *Safety Requirements for Plumbing*. Reprinted with permission.

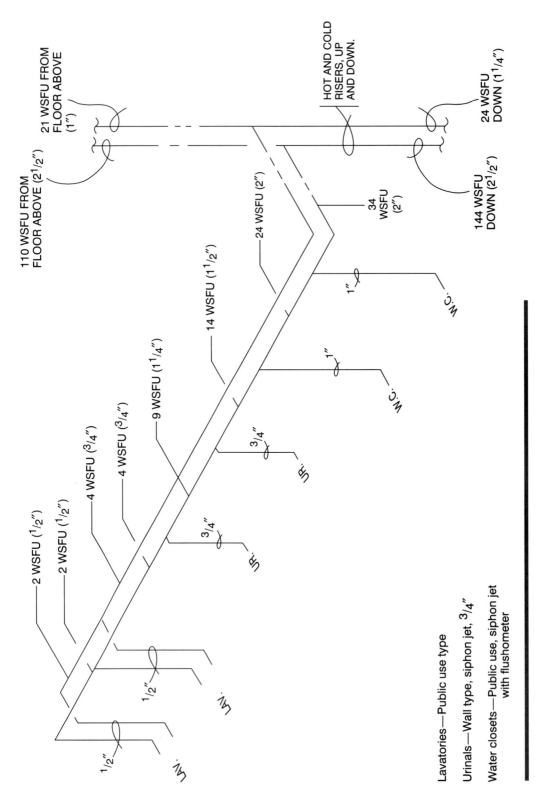

Figure 8–17 Domestic water sizing example.

SIZING A WATER BOOSTER PUMP

Knowing how to size a water pressure booster is useful for a construction manager because this process addresses the question of whether there is adequate water pressure at the highest or most distant plumbing fixture.

The process of designing a pressure booster system begins after the piping distribution has been designed and located on the plans. This process is simplified considerably by using Table 8–4. The steps to sizing a pressure booster system are as follows:

1. Calculate the static height of the system By "static height," we are referring to the total vertical distance from the city main to the highest plumbing fixture in the system. This, then, is the height to which we have to move the water. Remember that 1' of water is equal to 0.433 psi. Additionally, you will need to know the depth of bury of the city water main.

2. Determine the pressure required at the highest plumbing fixture We mentioned previously that certain plumbing fixtures require a specific pressure to work properly. For example, a tank-type water closet requires 15 psi of water pressure at the fixture to work properly.

3. Calculate the system friction loss Previously, we discussed the effects of pressure loss within the water distribution piping. The friction loss

Table 8–4 Sizing a Water Pressure Booster Pump

1. Static height (elevation (in ft) x 0.433 psig/ft)	= ____ psig
2. Pressure required at top	= ____ psig
3. Friction in piping system (ELR x AVG pressure drop/100 ft)	= ____ psig
4. Internal pump losses	= ____ psig
5. System pressure (Add steps 1, 2, 3, and 4)	= ____ psig
6. Minimum city pressure	= ____ psig
7. Meter loss	= ____ psig
8. Pressure drop (PD) for backflow preventer	= ____ psig
9. PD through softener	= ____ psig
10. PD through filters	= ____ psig
11. Minimum suction pressure (Add steps 7, 8, 9, and 10) Subtract from Step 6	= ____ psig
12. Required pump head (Boost) step 11 minus 5	= ____ psig

Table 1, reprinted with permission from *Plumbing Engineer* magazine.

in the system is calculated by multiplying the total developed length (TDL) of the system by the average pressure drop for the system.

The TDL of the piping system should be measured from the pump discharge to the most distant plumbing fixture and should include the pressure drop from valves and fittings, as found in Table 8–5. Adding the pressure drop for every valve and every fitting in a large system would be a very tedious task. Further, the plumbing contractor will undoubtedly install the system in a manner somewhat (or even a lot!) different from that shown on the plans in order to coordinate the installation of the plumbing with all other aspects of the construction (for example, it is usually necessary to relocate the plumbing lines to avoid conflicts with light fixtures, fire sprinkler piping, HVAC ductwork, and the structure of the building itself). For these reasons, most design engineers use an average to take care of the pressure losses in valves and fittings. Typically, for a simple system, many engineers will multiply the TDL by 150%, which usually takes care of the valve and fitting losses. If it is a complex system, with a lot of valves and fittings, the same engineer may use a 200% multiplier. Some engineers figure on 5% to 10% of the total static head. Either way, this gives a pretty close approximation of the total pressure drop within the system.

   4. **Assign a value for the pump losses**   If a booster pump is used, there will be some friction loss as the water travels through the pump itself. This is usually an assigned value or is provided by the pump manufacturer. This value is typically around 5 psi.

Adding Steps 1, 2, 3, and 4 will determine the total system pressure. This value should be entered as Item 5 on Table 8–4. If the required system pressure is less than the *lowest* pressure on the outlet of the backflow preventer, a pressure booster is not required.

The rest of the sizing steps are as follows:

   6. **Determine the *minimum* water pressure in the city main**   This requires a little research. Be alert when you are talking to individuals at the various public utilities. You are looking for the *lowest* pressure in the city main—not an average, not the "usual" pressure, and not "Well, as near as I can reckon."

   7. **Calculate the pressure losses through the water meter**   Most manufacturers of water meters can provide tables or graphs that define the amount of pressure drop as a function of the flow of water through a meter. Always use the data from the actual manufacturer of the water meter if at all possible.

   8. **Calculate the pressure losses through the backflow preventer**   Again, this information is best obtained from the manufacturer of the backflow preventer. This is usually a fairly significant number, often as high as 10 to 15 psi.

   9. **Calculate the pressure losses through the water softener**   This is assuming, of course, that there is a water softener. If there is, the pressure drop value should be obtained from the manufacturer if at all possible.

Table 8–5 Allowance in equivalent length of pipe for friction loss in valves and threaded fittings.

Diameter of Fitting (In.)	90° Std. Ell (ft)	45° Std. Ell (ft)	90° Side Tee (ft)	Coupling or Straight Run of Tee* (ft)	Gate Valve (ft)	Globe Valve (ft)	Angle Valve (ft)
3/8	1	0.6	1.5	0.3	0.2	8	4
1/2	2	1.2	3	0.6	0.4	15	8
3/4	2.5	1.5	4	0.8	0.5	20	12
1	3	1.8	5	0.9	0.6	25	15
1¼	4	2.4	6	1.2	0.8	35	18
1½	5	3	7	1.5	1.0	45	22
2	7	4	10	2	1.3	55	28
2½	8	5	12	2.5	1.6	65	34
3	10	6	15	3	2	80	40
3½	12	7	18	3.6	2.4	100	50
4	14	8	21	4.0	2.7	125	55
5	17	10	25	5	3.3	140	70
6	20	12	30	6	4	165	80

Table 2, reprinted with permission from *Plumbing Engineer* magazine.

10. **Calculate the pressure losses through the water filter** Some areas of the country, in addition to some well water systems, provide a filter on the domestic water line immediately after the water meter. If so, the pressure drop from the filter must be included in the calculations. As before, this information should come from the manufacturer.

The minimum suction pressure on the pump is simply Item 6 (the minimum pressure in the city main), less Items 7, 8, 9, and 10. The required "boost," or pump head, is the minimum suction pressure (Item 11) minus the system pressure (Item 5).

DOMESTIC WATER HEATING FUNDAMENTALS

Although there are numerous ways of heating domestic cold water to provide domestic hot water, the method used to classify this process is fairly simple. Basically, water heaters are classified by their fuel source (natural gas, propane, fuel oil, steam, or electric) and their storage capacity, if any. Because of this classification system, water heaters are usually technically described by their output in gallons per hour (gph) and their storage capacity. In the industry, the terms "recovery" and "storage" are commonly used.

Recovery is the heating capacity of the water heater, usually expressed in gallons per hour. Predictably, the recovery of the water heater is a function of the temperature rise that the water heater is trying to achieve. The higher the temperature rise, the lower the recovery from the same water heater. For example, Figure 8–19 shows that the recovery rate decreases as the temperature rise increases. This seems logical. After all, it takes more energy to increase one gallon of water 100° than it does to increase one gallon of water 50°.

The temperature differential used in sizing and defining a domestic water heater is the difference between the cold water coming into the water heater and the hot water leaving the water heater. For most residential and light commercial applications, the cold water incoming temperature is approximately the temperature of the city main. Depending on the source of the city water (wells versus reservoirs, for example) the cold water inlet temperature could range from 55°F to 78°F. In larger commercial applications, it could well be that the facility has some excess heat source and is able to preheat the cold water entering the water heater. In that case, the incoming temperature may be quite a bit higher than usual.

It stands to reason that recovery and storage have an inverse relationship; the higher the recovery rate, the smaller the storage tank can be. In other words, we can design or specify a water heater so that the output of the

A.O. SMITH

LIME TAMER™ MODELS

CONSERVATIONIST®
TANK-TYPE WATER HEATERS
BT-65, 80 & 100

FEATURES

All models meet ASHRAE/IES 90.1b-1992.

GLASS-LINED TANK — Assures years of rust-free clean hot water.

FULLY AUTOMATIC CONTROLS WITH SAFETY SHUTOFF — Accurate dependable control system requires no electric connections. Fixed automatic gas shutoff device for added safety. Not suggested for 180°F sanitizing.

HEAVY GAUGE STEEL JACKET — Finished with baked enamel over bonderized undercoat.

FOAM INSULATION — Saves fuel, helps reduce standby heat loss.

CERTIFICATION — Units are design certified by the American Gas Association (Canadian Gas Association for units built in Canada). Meets rigid requirements of the National Sanitation Foundation when equipped with leg kit. Certified for installation on combustible flooring.

EASY TO INSTALL — Completely factory assembled. Only gas, water and vent connections need be made. All connections are located in front and top of heaters for ease of installation and service.

DRAFT DIVERTER — Low profile diverter furnished as standard equipment.

MAXIMUM WORKING PRESSURE — 150 psi.

MAXIMUM GAS INLET PRESSURE — 14" W.C.

HANDHOLE CLEANOUT — On 75 and 100 gallon models. Allows easy tank cleaning.

OTHER FEATURES

- Built-in gas filter and integral dirt leg (propane only) • Anodic protection • Equipped with gas pressure regulator • Integral automatic gas shutoff system prevents excessive water temperature
- Factory installed A.G.A./ASME rated temperature and pressure relief valve.

LIMITED WARRANTY OUTLINE

If the tank should leak any time during the first three years, under the terms of the warranty, A. O. Smith will furnish a replacement heater; installation, labor, handling and local delivery extra. **THIS OUTLINE IS <u>NOT</u> A WARRANTY.** For complete information, consult the written warranty or A. O. Smith Water Products Company.

FOR UNITS BUILT IN USA

RECOVERY CAPACITIES

Model	Approx. Gal. Cap.	Type of Gas	Input Rating BTU/Hr.	Temperature Rise - Degrees F - Gallons Per Hour										
				30	40	50	60	70	80	90	100	110	120	130	140
* BT-65	50	Nat. & Prop.	50,000	162	131	97	81	69	61	54	48	44	40	37	35
BT-80	74.5	Nat. & Prop.	75,000	227	170	136	114	97	85	76	68	62	57	52	49
BT-100	100	Nat. & Prop.	75,000	227	170	136	114	97	85	76	68	62	57	52	49

* Recovery of BT-65 models based on 80% thermal efficiency.

NOTE: To compensate for the effects of high altitude areas above 2000 feet, recovery capacity should be reduced approximately 4% for every 1000 feet above sea level.

Capacity ratings are at 75% thermal efficiency (except as noted).

Revised December 1993   A 103.0

Figure 8–18 Sample tank-type water heater brochure.

heater (in gph) exactly matches the demand of the building (recovery equals demand). Therefore, if we need 162 gph of hot water, we simply make 162 gph of hot water, and we do not need to store any hot water. If we make less than 162 gph, we will need to store a certain percentage of the hot water to make up for the shortfall of hot water that will occur at peak loads. The selection of a hot water heater takes this relationship into account.

As with any mechanical device, the differences between storage and recovery has its advantages and disadvantages. Storage tanks have several advantages, including the ability to handle peaks and to downsize significantly the burners, flues, combustion air ducts, and gas lines to the hot water heater. The disadvantages to a storage tank are the valuable floor space needed for it, the weight of the tank (water weighs 8.346 pounds per gallon), and the significant amount of energy necessary to maintain the tank at its temperature setting.

TANK-TYPE WATER HEATERS

A tank-type water heater is a device that heats hot water and stores some of it for future use. Virtually all residences use this type of water heater. If the water heater is gas fired, it will have a gas burner that burns fuel (natural gas, propane, or fuel oil) to release heat, which, in turn, is transferred to the water in the tank. Accordingly, these types of water heaters will have a flue to convey the products of combustion outside the building to the atmosphere and a gas regulating and safety valve to prevent unwanted gas from building up in the combustion chamber. Additionally, any fuel-burning appliance will require a source of oxygen in order to burn the fuel, which means a supply of combustion air is required. In most instances, the combustion air is provided from within the volume of the building; however, if that volume is not sufficient, fresh air must be brought in from outside the building to provide the needed oxygen.

Tank-type water heaters can also use either steam, hot water, or electricity to heat the water, in a noncombustion process. Both steam and hot water use a coil that is immersed in the water in the tank. Hot water or steam flows through the coil, gives up its heat to the water, and flows back out the coil. An electric water heater, on the other hand, uses an immersion-type electrical coil to provide the same function.

Two special notes need to be made concerning storage tanks in water heating systems. All of the water in a hot water storage tank is not usable water. By this, we mean that some of the water in a hot water storage tank will not be maintained at a temperature that is suitable for use in the domestic hot water distribution system. This is because hot water stratifies in layers.

Hot water, like hot air, rises, leaving the colder, more dense water on the bottom of the tank. This explains why the hot water supply pipe is always taken from the top of the tank, where the hottest water is. For a vertical, hot water storage tank, we generally assume approximately 70% of the total tank capacity is usable. Accordingly, a construction manager should be alert to shop drawings or specifications that do not clearly specify the difference between gross storage and usable storage.

The second special note involves the use of temperature and pressure relief valves. All tank-type water heaters, and all storage tanks, must have a pressure and temperature relief valve installed on the tank, both to meet codes and regulations and as a matter of good common sense. Believe it or not, a simple gas-fired water heater, just like the one in your home, has enormous explosive potential. Water, as it changes from a liquid into a gas, increases its volume on the order of 1700 times. If, for some reason, the burner should continue to add heat to the hot water in the tank to the point where the water begins to change into steam, the tank literally becomes a bomb. To prevent this condition, water heaters are required to have a temperature and pressure relief valve that will discharge water if the temperature, or the pressure, in the hot water heater reaches a set limit. On most residential water heaters, this relief valve is set at 150 psi and 210°F. Note also that water heaters are required to have a safety gas shut-off valve, which automatically shuts off the gas to the hot water heater in the event of a failure of the pilot light.

INSTANTANEOUS WATER HEATERS

Water heaters designed so that the recovery equals the demand do not need storage. These types of water heaters are referred to as instantaneous or semi-instantaneous water heaters.

As with tank-type water heaters, instantaneous heaters can utilize steam, hot water, natural gas, propane, fuel oil, or electricity to heat the water. Because they are sized to meet the entire domestic hot water demand of the building, they tend to be found in larger facilities, such as hospitals, factories, and large office complexes (see Figure 8–19).

Smaller versions of these larger commercial units are commonly used as booster heaters for commercial dishwashers (which require, by code, 180° water). Another similar type of water heater is a small electric "point of use" heater that fits underneath a sink or within a vanity and serves only one or two lavatories. These small instantaneous heaters work well where a few isolated fixtures are located a long distance from the hot water heater and minimal storage is required.

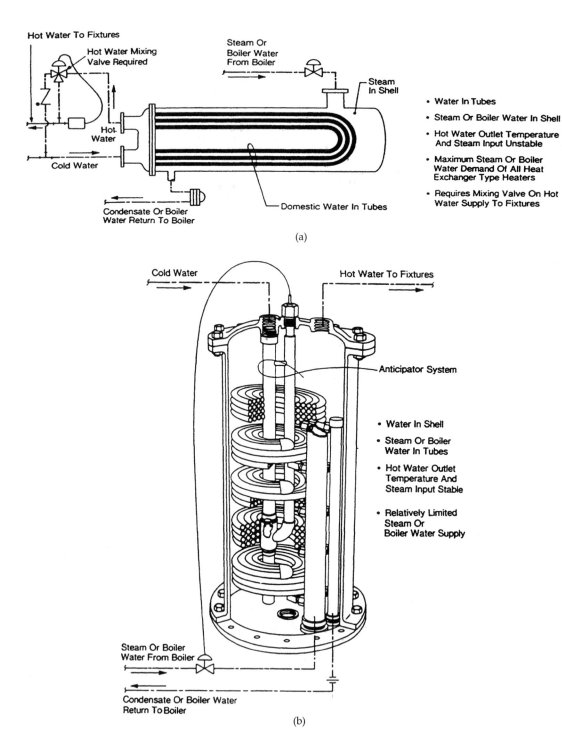

Figure 8–19 (a) Instantaneous water heater; (b) semi-instantaneous water heater.
Courtesy *Plumbing Engineer* magazine.

MULTIPLE TEMPERATURE APPLICATIONS

In many commercial applications, there is a need for domestic hot water at a variety of temperatures. In a hospital, for example, there may be a need for 180°F hot water for the dishwasher, 140°F hot water for general kitchen use, 120°F water for cleaning and maintenance, and 105°F hot water for showers and hand washing. Additionally, each of these lines may have its own recirculation line. In other words, it would not be unusual for a hospital to have eight domestic hot water lines, at different temperatures, serving the main hospital!

It would be very expensive to have a separate water heater for each system, so many such institutions use dual or even triple temperature water heaters. Typically, these systems use a water heater that produces hot water at the highest temperature required, then uses a series of mixing valves to create a second, or third, supply of water at a lower temperature (see Figure 8–20).

These types of systems, if designed and installed properly, work very well and are far more cost effective than installing separate water heaters for each system.

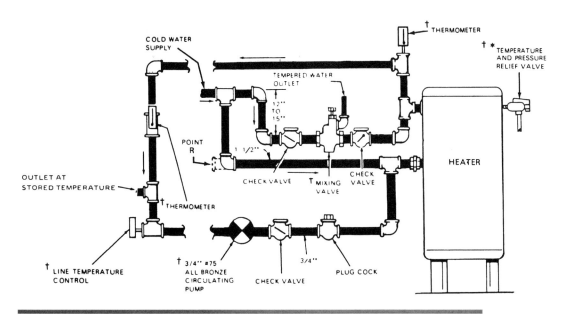

Figure 8–20 Heater with mixing valve and recirculated loop.
Courtesy A. O. Smith Water Products Corp.

SIZING A HOT WATER HEATER

The three main questions to answer when sizing a water heater are: How much hot water do we need?, What temperature does the water need to be?, and What length of time will I have to recover this demand? Once you reach an understanding of these three questions for the particular building under study, the selection of a hot water heater becomes relatively simple.

As with most procedures in the plumbing industry, the process of sizing a domestic water heater is usually done by computer. For the purposes of this text, we will be using the "Acc-U-Size" commercial water heater sizing program offered by A. O. Smith Water Products. The steps to sizing a domestic hot water heater, using this program, are as follows:

1. **Select the building type** There are three choices of building type: food service, multiple dwelling, and other water heater applications. The multiple dwelling dialogue box is further subdivided into apartments, hotels/motels, and dormitories. The other applications are subdivided into a multitude of applications including office buildings, country clubs, beauty shops, and many other applications.

2. **Select the fuel source** The options given are natural gas, liquid propane (LP), electric, and oil.

3. **Select the primary heater application** This dialogue box gives you the opportunity to select a single tank heater, a tank heater with a storage tank, or multiple tank heaters, with or without storage tanks.

4. **Select heater inlet and outlet temperatures** This procedure was discussed previously in the text.

5. **Select quantities of fixtures, people, or area** Each of the three dialogue boxes has different methods of ascertaining the demand of hot water for the application that you have selected. For example, an office building application is based on either the number of people in the office building or the gross area of the office building. On the other hand, the sizing for a hotel/motel is based on either the number of rooms or the number of people. Many of these dialogue boxes also require information about the type of shower head or faucet used (low flow vs. standard flow) and the peak recovery period in hours. There is also an entry point if you have some other loads at a differing temperature, for example, 140°F hot water to a kitchen.

6. **Select whether or not you want ASME design** Some states require the vessel of a water heater to be stamped with an ASME (American Society of Mechanical Engineers) assembly stamp, if the hot water heater is of a certain size or larger. This is important to know as the extra quality control required by an ASME stamp will significantly increase the price of the hot water heater.

7. Activate the "Calculate Design" button This produces a dialogue box that will give you a number of possible correct choices, usually with one or two marked as "recommended." Highlighting the water heater brings up a dialogue box that will identify the water heater(s) and storage tanks selected and allow you to print the specifications, or the operating costs, for this particular application.

REVIEW QUESTIONS

1. What are the four special considerations to keep in mind when designing a domestic hot water system?
2. Describe water hammer and how it is created.
3. What is the maximum allowable working pressure (MAWP) of a domestic water system? Why is this important?
4. What is the maximum allowable water velocity within a domestic water system? Why is this important?
5. What is the purpose of the domestic hot water recirculation system? Why is it necessary?
6. What is the purpose and function of the backflow preventer installed on the water service immediately upon entering the building?
7. What is the difference between an instantaneous water heater and a tank-type water heater?
8. Draw an isometric diagram of the following bathroom group and size all piping.

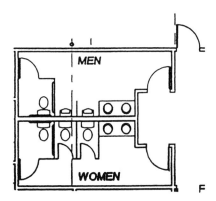

9. Draw an isometric diagram for the following bathroom group and size all piping.

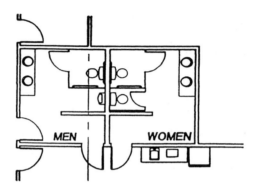

10. A four-story building has the following design parameters. Determine whether this particular building requires a booster pump and, if so, how much boost is required.
    a. Static height of the system is 50′.
    b. The developed length of the piping, less the losses for the fittings and valves, is 195′. The water distribution system is a simple system (that is, you have to add the losses for the fittings).
    c. Assume any internal pump losses would be 5 psi.
    d. The city main pressure is 75 psi, plus or minus 5 psi.
    e. The meter loss is 8 psi.
    f. The 2″ backflow preventer has a pressure drop of 25′ of head.
    g. There is no softener or water filter in the domestic water system.
    h. The main entering the building is 2″ in size.
    i. The water closets have flushometers and are the highest fixture in the building.

APPENDIX

Basic Principles*

The objective of this Code is to implement basic principles of environmental sanitation and safety through properly designed, acceptably installed, and adequately maintained plumbing systems. The basic principles enumerated below presuppose the normal and proper use of the plumbing system according to its design and purpose.

Approval of a material or design for purposes of this Code, as limited and qualified in the Code, means that a national consensus of informed opinion has judged the material or design to be at least minimally adequate for protection of the public health and safety, and the economic interest of the consumer. Such judgement is supported by the approving action of the American National Standards Institute.

As interpretations may be required, and as unforeseen situations arise which are not specifically covered in this Code, the principles which follow shall be used to define the intent.

Principle 1—Consumer Protection

The keystone of all good plumbing is consumer protection; protection not only of the consumer's health and safety, but of the consumer's financial invest-

---

Basic Principles, ANSI A40–1993 Standard Safety Requirements for Plumbing, reprinted with permission.

ment as well. A primary objective of this Code is that each component of the plumbing system shall be durable material free from defective workmanship and designed and constructed to give satisfactory service for its reasonable life. Under normal usage, non-accessible components of the plumbing system manufactured in compliance with standards listed in Table 1, when properly designed, installed, and maintained, should have a useful life of 20 years.

Principle 2—Potable Water

All premises intended for human habitation, occupancy, or use shall be provided with a supply of potable water. Such a water supply shall not be directly connected to unsafe water sources and shall be protected from the hazards of backflow and back siphonage.

Principle 3—Hot Water

Hot water shall be supplied for every dwelling for human habitation.

Principle 4—Plumbing Fixtures

Each family dwelling unit shall have at least one water closet, one lavatory, one kitchen-type sink, and one bathtub or shower to meet the basic requirements of sanitation and personal hygiene. All other structures for human habitation shall be equipped with sanitary facilities in accordance with this Code. The plumbing fixtures provided shall be made of durable, smooth, non-absorbent, and corrosion-resisting material and shall be free from concealed fouling surfaces.

Principle 5—Structural Safety

Plumbing shall be installed with due regard to preservation of the strength of structural members and prevention of damage to walls and other surfaces from fixture usage.

Principle 6—Safety Devices

Devices for heating and storing water shall be so designed and installed as to guard against dangers from explosion or overheating.

Principle 7—Ground and Surface Water Protection

Sewage or other waste shall not be discharged into surface or subsurface water unless it has first been subjected to an approved form of treatment.

Principle 8—Water Conservation

Plumbing shall be designed and adjusted to use the minimum quantity of water consistent with proper performance and cleaning.

Principle 9—Public Sewer and Water

Every building intended for human habitation, occupancy, or use, and located near a public sewer and water supply, shall be connected to said sewer and water supply, respectively, or both where available.

Principle 10—Sewer Flooding Prevention

Where a plumbing drainage system is subject to backflow of sewage from the public sewer, suitable provision shall be made to prevent such backflow into the building.

Principle 11—Vent Terminal

Each vent terminal shall extend to the outer air and be so installed as to minimize the possibilities of clogging and the return of foul air to the building.

Principle 12—Drainage System

The drainage system shall be designed and installed to prevent fouling, deposit of solids, and clogging, and shall be provided with cleanouts so arranged that they may be readily cleaned and maintained as required by this Code.

Principle 13—Pressure and Volume

Plumbing fixtures, devices, and appurtenances shall be supplied with water in sufficient volume and at adequate pressures to enable them to function properly under normal conditions of use.

Principle 14—Fixture Traps

Each plumbing fixture directly connected to the drainage system shall be equipped with a liquid seal trap.

Principle 15—Trap Seals Protection

The drainage system shall be provided with ventilation to guard against self-siphonage, induced siphonage, or forcing of trap seals under normal conditions of use as specified in this Code.

Principle 16—Light and Ventilation

No water closet or similar plumbing fixture shall be installed in a room or compartment which is not lighted or ventilated.

Principle 17—Fixture Accessibility

All plumbing fixtures shall be so installed with regard to spacing as to be accessible for their intended use, cleaning, and maintenance.

Principle 18—Private Sewage Disposal Systems

When plumbing fixtures are installed in a building having no public sanitary sewer available, provision shall be made for private sewage disposal by an approved method.

Principle 19—Tests

The plumbing system shall be subjected to such tests as will disclose any leaks and defects in the work or the material. Tests on innovative systems shall be made to demonstrate adequacy of performance regarding sanitary function, health, and safety.

Principle 20—Maintenance

Plumbing systems shall be so designed and installed as to facilitate repair and maintenance, and shall be maintained in a safe and serviceable condition from the standpoint of both operation and health.

APPENDIX

Definitions

Air Break (Drainage System)

A piping arrangement in which a drain from a fixture, appliance, or device discharges indirectly into a fixture, receptor, or interceptor at a point below the flood level rim of the receptor. (See Figure 1.2.2.)

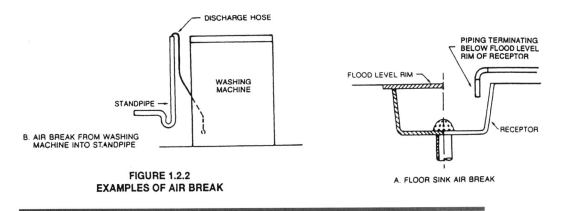

Figure 1.2.2 Examples of air break.

All illustrations are reprinted with permission from the *National Standard Plumbing Code*, 1993 Edition, published by the National Association of Plumbing Heating Cooling Contractors.

Air Chamber

A pressure surge absorbing device operating through the compressibility of air. (See Figure 1.2.3.)

Figure 1.2.3 One type of air chamber.

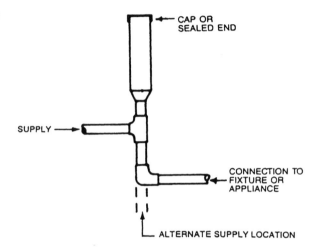

FIGURE 1.2.3 ONE TYPE OF AIR CHAMBER

Air Gap (Drainage System)

The unobstructed vertical distance through the free atmosphere between the outlet of the waste pipe and the flood level rim of the receptor into which it is discharging. (See Figure 1.2.4.)

Figure 1.2.4 An example of air gap (drainage system).

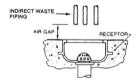

FIGURE 1.2.4 AN EXAMPLE OF AIR GAP (DRAINAGE SYSTEM)

Air Gap (Water Distribution System)

The unobstructed vertical distance through the free atmosphere between the lowest opening from any pipe or faucet supplying water to a tank, plumbing fixture, or other device and the flood level rim of the receptor. (See Figure 1.2.5.)

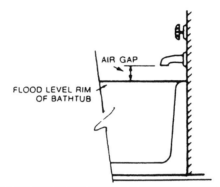

FIGURE 1.2.5 AN EXAMPLE OF AIR GAP (WATER DISTRIBUTION SYSTEM)

Figure 1.2.5 An example of air gap (water distribution system).

Area Drain

A receptor designed to collect surface or storm water from an open area. (See Figure 1.2.6.)

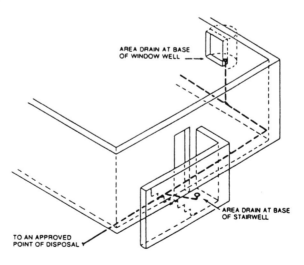

FIGURE 1.2.6 TYPICAL AREA AND WINDOW WELL DRAIN INSTALLATIONS

Figure 1.2.6 Typical area and window well drain installations.

Backflow Preventer

A device or means to prevent backflow. (See Figures 1.2.25 and 1.2.45.)

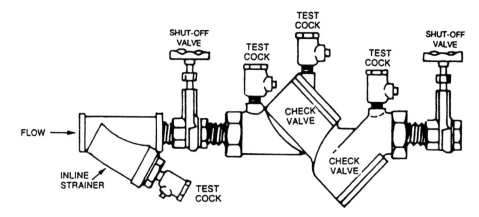

FIGURE 1.2.25 AN EXAMPLE OF A DOUBLE CHECK VALVE ASSEMBLY

Figure 1.2.25 An example of a double-check valve assembly.

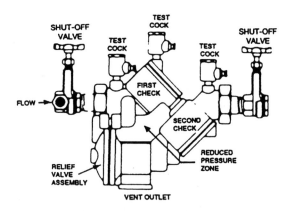

FIGURE 1.2.45 AN EXAMPLE OF REDUCED PRESSURE PRINCIPLE BACKPRESSURE BACKFLOW PREVENTER—FLOW POSITION

Figure 1.2.45 An example of reduced pressure principle backpressure backflow preventer—flow position.

Backflow—Water Distribution

The flow of water or other liquids, mixtures, or substances into the distributing pipes of a potable supply of water from any source or sources other than its intended source. Back-siphonage is one type of backflow.

Backpressure Backflow

A condition which may occur in the potable water distribution system, whereby a higher pressure than the supply pressure is created, causing a reversal of flow into the potable water piping. (See Figure 1.2.8.)

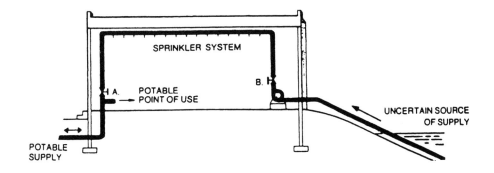

A. **CONTACT POINT:** A SINGLE VALVED CONNECTION (A) EXISTS BETWEEN THE PUBLIC POTABLE WATER SUPPLY AND A SOURCE OF WATER SUPPLY OF UNCERTAIN QUALITY

B. **CAUSE OF REVERSED FLOW:** THE SPRINKLER SYSTEM IS NORMALLY SUPPLIED FROM A NEARBY SOURCE OF UNCERTAIN SUPPLY THROUGH A HIGH PRESSURE PUMP. WHEN THE VALVE (A) IS LEFT OPEN CONTAMINATED LAKE WATER CAN BE PUMPED TO THE PUBLIC SUPPLY

C. **SUGGESTED CORRECTION:** THE POTABLE WATER SUPPLY TO THE SPRINKLER SYSTEM SHOULD BE PROTECTED ACCORDING TO SECTIONS 10.5 AND 10.5.9

FIGURE 1.2.8 AN EXAMPLE OF BACKFLOW CAUSED BY BACKPRESSURE

Figure 1.2.8 An example of backflow caused by backpressure.

Back-Siphonage

The flowing back of used, contaminated, or polluted water from a plumbing fixture or vessel or other sources into a potable water supply pipe due to a negative pressure in such pipe. (See Figure 1.2.9.)

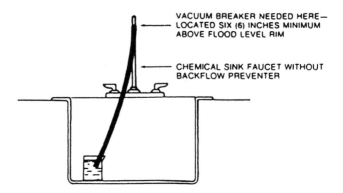

A. **CONTACT POINT:** RUBBER HOSE ON CHEMICAL FAUCET IS SUBMERGED IN SOLUTION WHICH CONTAINS TOXIC CHEMICALS

B. **CAUSE OF REVERSED FLOW:** NEGATIVE PRESSURE (VACUUM) ON POTABLE SUPPLY CAUSING CHEMICALS TO ENTER POTABLE WATER PIPING

C. **SUGGESTED CORRECTION:** ATMOSPHERIC VACUUM BREAKER ON FAUCET OUTLET

FIGURE 1.2.9 BACKFLOW CAUSED BY BACK SIPHONAGE

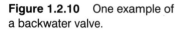

Figure 1.2.9   Backflow caused by back siphonage.

Back Water Valve

A device installed in a drain or pipe to prevent backflow. (See Figure 1.2.10.)

Figure 1.2.10   One example of a backwater valve.

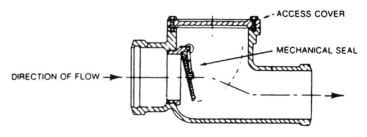

FIGURE 1.2.10 ONE EXAMPLE OF A BACKWATER VALVE

Branch

Any part of the piping system other than a riser, main or stack. (See Figure 1.2.13.)

FIGURE 1.2.13 EXAMPLES OF BRANCHES

Figure 1.2.13 Examples of branches.

Building Drain

That part of the lowest piping of a drainage system which receives the discharge from soil, waste, and other drainage pipes inside the walls of the building and conveys it to the building sewer beginning 3' outside the building wall.

Building Sewer

That part of the drainage system which extends from the end of the building drain and conveys its discharge to a public sewer, private sewer, individual sewage-disposal system, or other points of disposal.

Building Sewer—Combined

A building sewer that conveys both sewage and storm water or other drainage. (See Figure 1.2.15.)

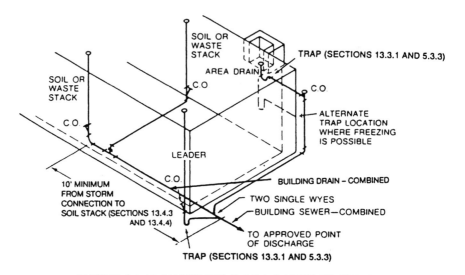

FIGURE 1.2.15 AN EXAMPLE OF A COMBINED SYSTEM

Figure 1.2.15 An example of a combined system.

Building Subdrain

That portion of a drainage system that does not drain by gravity into the building sewer. (See Figure 1.2.17.)

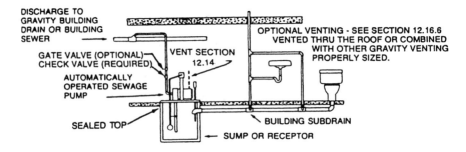

FIGURE 1.2.17 AN EXAMPLE OF A BUILDING SUBDRAIN

Figure 1.2.17 An example of a building subdrain.

Building Trap

A device, fitting, or assembly of fittings, installed in the building drain to prevent circulation of air between the drainage system of the building and the building sewer. (See Figure 1.2.18.)

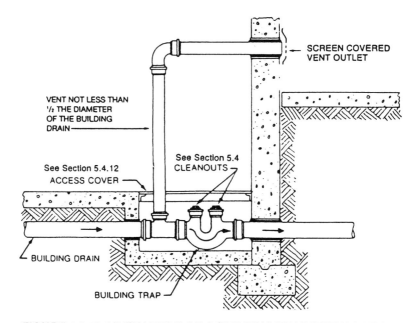

FIGURE 1.2.18 AN EXAMPLE OF A BUILDING TRAP INSTALLATION

Figure 1.2.18 An example of a building trap installation.

Cesspool

A lined and covered excavation in the ground that receives the discharge of domestic sewage or other organic wastes from a drainage system, designed to retain the organic matter and solids, but permitting the liquids to seep through the bottom and sides.

Combination Fixture

A fixture combining one sink and laundry tray or a two- or three-compartment sink or laundry tray in one unit.

Combination Waste and Vent System

A specifically designed system of waste piping embodying the horizontal wet venting of one or more sinks or floor drains by means of common waste and vent pipe adequately sized to provide free movement of air above the flow line of the drain. (See Figure 1.2.20.)

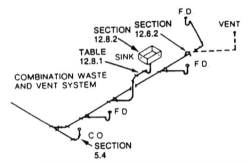

FIGURE 1.2.20 AN EXAMPLE OF COMBINATION WASTE AND VENT SYSTEM - SIZED AS PER SECTION 12.17

Figure 1.2.20 An example of combination waste and vent system.

Conductor

The water conductor from the roof to the building storm drain, combined building sewer, or other means of disposal, and located inside the building.

Continuous Waste

A drain from two or more fixtures connected to a single trap. (See Figure 1.2.22.)

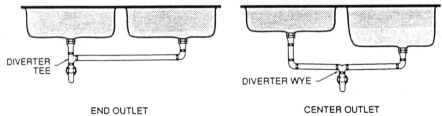

FIGURE 1.2.22 EXAMPLE OF CONTINUOUS WASTE ARRANGEMENTS

Figure 1.2.22 Example of continuous waste arrangements.

Cross Connection

Any connection or arrangement between two otherwise separate piping systems, one of which contains potable water and the other either water of questionable safety, steam, gas, or chemical whereby there may be a flow from one system to the other, the direction of flow depending on the pressure differential between the two systems. (See Figures 1.2.8 and 1.2.9.)

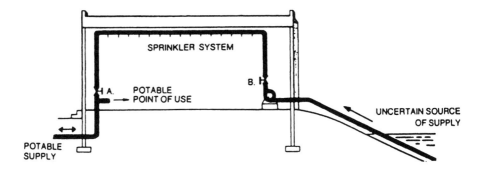

A. **CONTACT POINT:** A SINGLE VALVED CONNECTION (A) EXISTS BETWEEN THE PUBLIC POTABLE WATER SUPPLY AND A SOURCE OF WATER SUPPLY OF UNCERTAIN QUALITY

B. **CAUSE OF REVERSED FLOW:** THE SPRINKLER SYSTEM IS NORMALLY SUPPLIED FROM A NEARBY SOURCE OF UNCERTAIN SUPPLY THROUGH A HIGH PRESSURE PUMP. WHEN THE VALVE (A) IS LEFT OPEN CONTAMINATED LAKE WATER CAN BE PUMPED TO THE PUBLIC SUPPLY

C. **SUGGESTED CORRECTION:** THE POTABLE WATER SUPPLY TO THE SPRINKLER SYSTEM SHOULD BE PROTECTED ACCORDING TO SECTIONS 10.5 AND 10.5.9

FIGURE 1.2.8 AN EXAMPLE OF BACKFLOW CAUSED BY BACKPRESSURE

Figure 1.2.8 An example of backflow caused by backpressure.

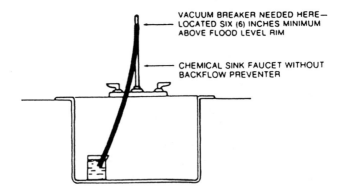

A. **CONTACT POINT:** RUBBER HOSE ON CHEMICAL FAUCET IS SUBMERGED IN SOLUTION WHICH CONTAINS TOXIC CHEMICALS

B. **CAUSE OF REVERSED FLOW:** NEGATIVE PRESSURE (VACUUM) ON POTABLE SUPPLY CAUSING CHEMICALS TO ENTER POTABLE WATER PIPING

C. **SUGGESTED CORRECTION:** ATMOSPHERIC VACUUM BREAKER ON FAUCET OUTLET

FIGURE 1.2.9 BACKFLOW CAUSED BY BACK SIPHONAGE

Figure 1.2.9   Backflow caused by back siphonage.

Double-Check Valve Assembly

A backflow prevention device consisting of two independently acting check valves, internally force loaded to a normally closed position between two tightly closing shut-off valves, and with means of testing for tightness. (See Figure 1.2.25.)

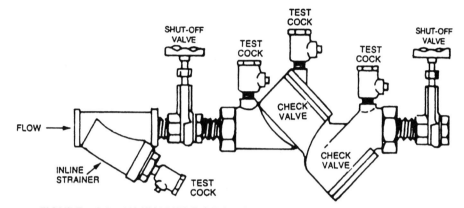

FIGURE 1.2.25 AN EXAMPLE OF A DOUBLE CHECK VALVE ASSEMBLY

Figure 1.2.25 An example of a double-check valve assembly.

Drain

Any pipe that carries waste or water-borne wastes in a building drainage system.

Drainage System

Includes all the piping, within public or private premises, that conveys sewage, rainwater, or other liquid wastes to a point of disposal. It does not include the mains of a public sewer system or private or public sewage treatment.

DWV

An acronym for "drain-waste-vent," referring to the combined sanitary drainage and venting systems. This term is technically equivalent to "soil-waste-vent" (SWV).

Effective Opening

The minimum cross-sectional area at the point of water supply discharge, measured, or expressed in terms of (1) diameter of a circle, or (2) if the opening is not circular, the diameter of a circle of equivalent cross-sectional area. (See Figure 1.2.26.)

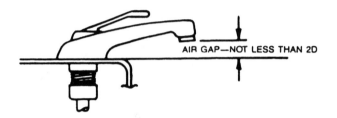

FIGURE 1.2.26a FAUCET SPOUT EFFECTIVE OPENING

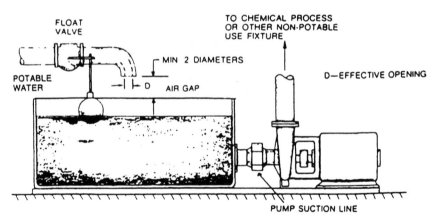

FIGURE 1.2.26b AN EXAMPLE OF AN EFFECTIVE OPENING AND AIR GAP SIZING

Figure 1.2.26 (a) Faucet spout effective opening; (b) an example of an effective opening and air gap sizing.

Equivalent Length

The length of straight pipe of a specific diameter that would produce the same frictional resistances as a particular fitting or line comprised of pipe and fittings.

Fixture Branch—Drainage

A drain serving one or more fixtures which discharges into another drain. (See Figure 1.2.27.)

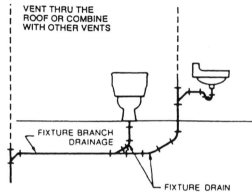

FIGURE 1.2.27 EXAMPLES OF FIXTURE DRAIN AND FIXTURE BRANCH, DRAINAGE

Figure 1.2.27 Examples of fixture drain and fixture branch drainage.

Fixture Drain

The drain from the trap of a fixture to the junction of that drain with any other drain pipe. (See Figure 1.2.27.)

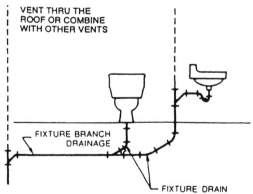

FIGURE 1.2.27 EXAMPLES OF FIXTURE DRAIN AND FIXTURE BRANCH, DRAINAGE

Figure 1.2.27 Examples of fixture drain and fixture branch drainage.

Fixture Supply

The water supply pipe connecting a fixture, a branch water supply pipe, or directly to a main water supply pipe. (See Figure 1.2.28.)

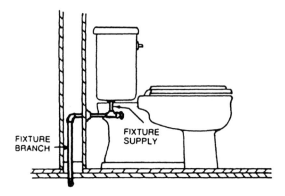

FIGURE 1.2.28 AN EXAMPLE OF A FIXTURE SUPPLY AND FIXTURE BRANCH

Figure 1.2.28   An example of a fixture supply and fixture branch.

Fixture Unit (Drainage—DFU)

A measure of the probable discharge into the drainage system by various types of plumbing fixtures. The drainage fixture-unit value for a particular fixture depends on its volume rate of drainage discharge, on the time duration of a single drainage operation, and on the average time between successive operations.

Fixture Unit (Supply—SFU)

A measure of the probable hydraulic demand on the water supply by various types of plumbing fixtures. The supply fixture-unit value for a particular fixture depends on its volume rate of supply, on the time duration of a single supply operation, and on the average time between successive operations.

Flood Level Rim

The edge of the receptor from which water overflows. (See Figure 1.2.29.)

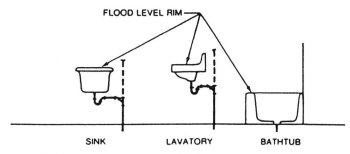

FIGURE 1.2.29 EXAMPLES OF FLOOD LEVEL RIM

Figure 1.2.29 Examples of flood level rim.

Flow Pressure

The pressure in the water supply pipe near the faucet or water outlet while the faucet or water outlet is fully open and flowing. (See Figure 1.2.30.)

Figure 1.2.30 An example of flow pressure.

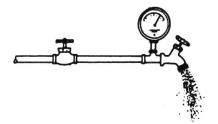

FIGURE 1.2.30 AN EXAMPLE OF FLOW PRESSURE

Flush Valve

A device located at the bottom of a tank for flushing water closets and similar fixtures. (See Figure 1.2.32.)

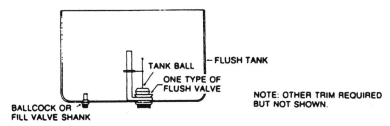

FIGURE 1.2.32 AN EXAMPLE OF FLUSH VALVE IN FLUSH TANK

Figure 1.2.32 An example of flush valve in flush tank.

Flushometer Tank

A device integrated within an air accumulator vessel which is designed to discharge a predetermined quantity of water to the fixture for flushing purposes. (See Figure 1.2.32.)

Figure 1.2.32 An example of a flushometer tank.

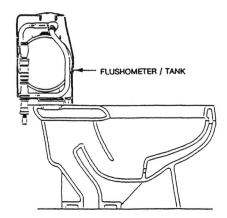

FIGURE 1.2.32a AN EXAMPLE OF FLUSHOMETER TANK

Flushometer Valve

A device that discharges a predetermined quantity of water to fixtures for flushing purposes and is closed by direct water pressure or other mechanical means. (See Figure 1.2.33.)

Figure 1.2.33

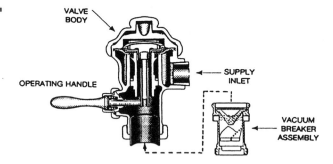

Grade

The fall (slope) of a line of pipe in reference to a horizontal plane. In drainage, it is usually expressed as the fall in a fraction of an inch per foot length of pipe. (See Figure 1.2.34.)

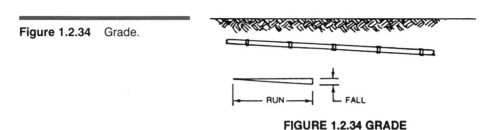

Figure 1.2.34 Grade.

Horizontal Branch Drain

A drain branch pipe extending laterally from a soil or waste stack or building drain, with or without vertical sections or branches, that receives the discharge from one or more fixture drains and conducts it to the soil or waste stack or to the building drain.

Hot Water

Hot water is supplied to plumbing fixtures at a temperature of not less than 120° F and not more than 140°F except that commercial dishwashing machines and similar equipment shall be provided with water 180°F for sterilization purposes.

Indirect Waste Pipe

A waste pipe that does not connect directly with the drainage system but discharges into the drainage system through an air break or air gap into a trap, fixture, receptor, or interceptor. (See Figure 1.2.36.)

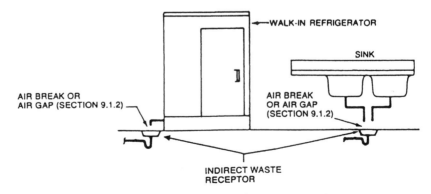

Figure 1.2.36 Indirect waste pipe.

Interceptor

A device designed and installed so as to separate and retain deleterious, hazardous, or undesirable matter from normal wastes while permitting normal sewage or liquid wastes to discharge into the drainage system by gravity. (See Figure 1.2.37.)

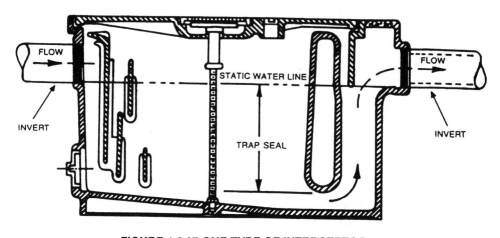

FIGURE 1.2.37 ONE TYPE OF INTERCEPTOR

Figure 1.2.37 One type of interceptor.

Invert

The lowest portion of the inside of a horizontal pipe. (See Figure 1.2.37a.)

Figure 1.2.37a

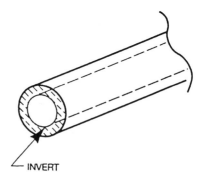

FIGURE 1.2.37a

Leaching Well or Pit

A pit or receptor having porous walls which permit the contents to seep into the ground. (See Figure 1.2.38.)

Figure 1.2.38 An example of a leaching well or pit.

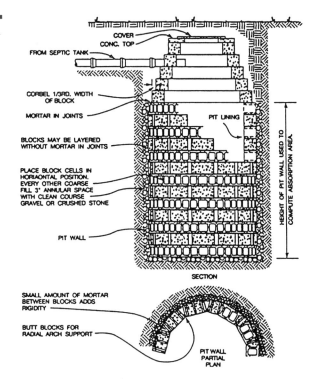

FIGURE 1.2.38 AN EXAMPLE OF A LEACHING WELL OR PIT

Leader

An exterior vertical drainage pipe for conveying storm water from roof or gutter drains. (See Figure 1.2.39.)

Figure 1.2.39 An example of a leader.

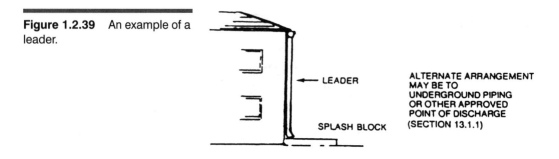

FIGURE 1.2.39 AN EXAMPLE OF LEADER

Main

The principal pipe artery to which branches may be connected.

Offset

A combination of elbows or bends that brings one section of the pipe out of line but into a line parallel with the other section. (See Figure 1.2.41.)

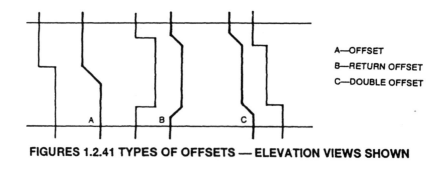

FIGURES 1.2.41 TYPES OF OFFSETS — ELEVATION VIEWS SHOWN

Figure 1.2.41 Types of offsets—elevation views shown.

Plumbing Fixture

A receptacle or device that is either permanently or temporarily connected to the water distribution system of the premises and demands a supply of water therefrom; that discharges used water, liquid-borne waste materials, or sewage either directly or indirectly to the drainage system of the premises; or that requires both a water supply connection and a discharge to the drainage system of the premises. Plumbing appliances as a special class of fixture are further defined.

Note: Some examples of plumbing fixtures are water closets, lavatories, sinks, drinking fountains, urinals, and similar devices.

Potable Water

Water free from impurities present in amounts sufficient to cause disease or harmful physiological effects and conforming in its bacteriological and chemical quality to the requirements of the Public Health Service Drinking Water Standards or the regulations of the public health authority having jurisdiction.

Private Sewage Disposal System

A system for disposal of domestic sewage by means of a septic tank or mechanical treatment, designed for use apart from a public sewer to serve a single establishment or building. (See Figure 1.2.42.)

160 Definitions

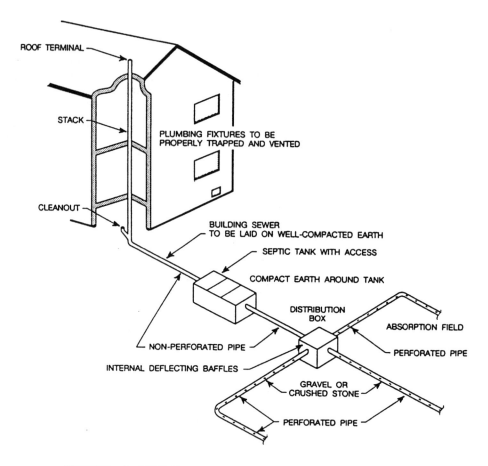

Figure 1.2.42 Typical private sewage disposal system.

Private Sewer

A sewer not directly controlled by public authority.

Private Water Supply

A supply other than an approved public water supply which serves one or more buildings. (See Figure 1.2.43.)

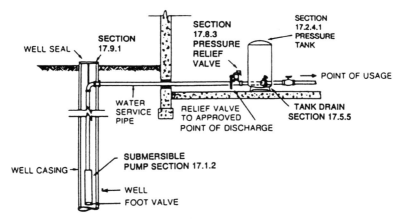

FIGURE 1.2.43 TYPICAL ARRANGEMENT OF PRIVATE WATER SUPPLY

Figure 1.2.43 Typical arrangement of a private water supply.

Reduced Pressure Principle Backpressure Backflow Preventer

A backflow prevention device consisting of two independently acting check valves, internally force loaded to a normally closed position and separated by an intermediate chamber (or zone) in which there is an automatic relief means of venting to the atmosphere, internally loaded to a normally open position between two tightly closing shut-off valves and with means for testing for tightness of the checks and opening of relief means. (See Figure 1.2.45.)

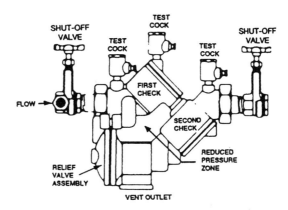

FIGURE 1.2.45 AN EXAMPLE OF REDUCED PRESSURE PRINCIPLE BACKPRESSURE BACKFLOW PREVENTER—FLOW POSITION

Figure 1.2.45 An example of reduced pressure principle backpressure backflow preventer—flow position.

Riser

A water supply pipe that extends vertically one full story or more to convey water to branches or to a group of fixtures. (See Figure 1.2.46.)

Figure 1.2.46 Riser to drinking fountains.

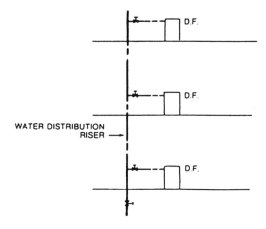

FIGURE 1.2.46 RISER TO DRINKING FOUNTAINS

Roof Drain

A drain installed to receive water collecting on the surface of a roof and to discharge it into a leader or a conductor. (See Figure 1.2.47.)

Figure 1.2.47 An example of a roof drain.

FIGURE 1.2.47 AN EXAMPLE OF A ROOF DRAIN

Roughing-In

The installation of all parts of the plumbing system that can be completed prior to the installation of fixtures. This includes drainage, water supply, and vent piping, and the necessary fixture supports, or any fixtures that are built into the structure.

Sand Filter

A treatment device or structure, constructed above or below the surface of the ground, for removing solid or colloidal material of a type that cannot be removed by sedimentation from septic tank effluent. (See Figure 1.2.48.)

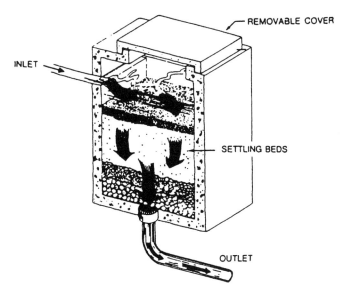

Figure 1.2.48 An example of a sand filter.

FIGURE 1.2.48 AN EXAMPLE OF A SAND FILTER

Septic Tank

A watertight receptacle that receives the discharge of a building sanitary drainage system or part thereof and is designed and constructed to separate solids from the liquid, digest organic matter through a period of detention, and allow the liquids to discharge into the soil outside of the tank through a system of open joint or perforated piping or a seepage pit. (See Figure 1.2.49.)

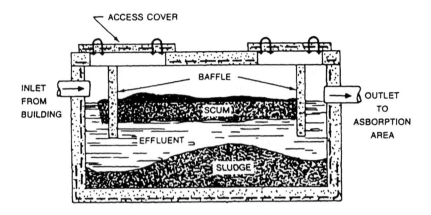

Figure 1.2.49 An example of a septic tank.

Sewage Ejectors—Pneumatic

A device for lifting sewage by air pressure. (See Figure 1.2.50.)

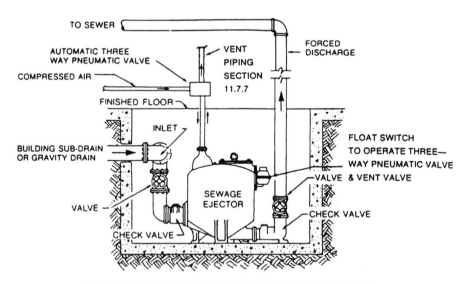

Figure 1.2.50 An example of a sewage ejector.

Sewage Pump

A permanently installed mechanical device other than an ejector for removing sewage or liquid waste from a sump. (See Figure 1.2.51.)

Figure 1.2.51 An example of a sewage pump.

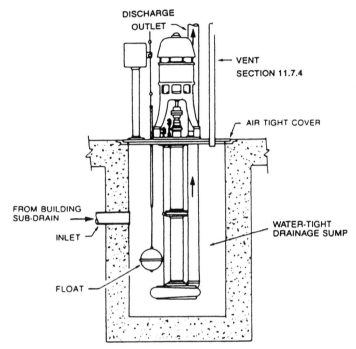

FIGURE 1.2.51 AN EXAMPLE OF SEWAGE PUMP

Special Wastes

Wastes that require special treatment before entry into the normal plumbing system. (See Figure 1.2.52.)

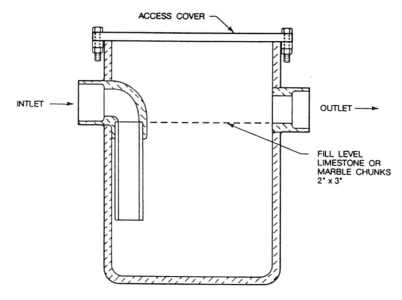

FIGURE 1.2.52 AN EXAMPLE OF SPECIAL WASTE NEUTRALIZING TANK

Figure 1.2.52 An example of a special waste-neutralizing tank.

Stack

A general term for any vertical line including offsets of soil, waste, vent, or inside conductor piping. This does not include vertical fixture and vent branches that do not extend through the roof or that pass through not more than two stories before being reconnected to the vent stack or stack vent. (See Figure 1.2.53.)

Figure 1.2.53 Stack and stack terms.

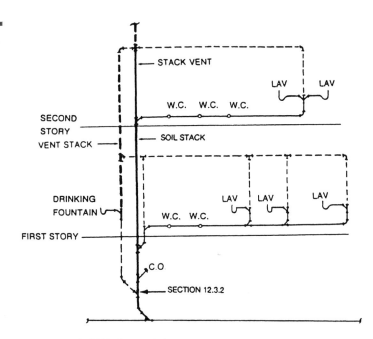

FIGURE 1.2.53 STACK AND STACK TERMS

Stack Vent

The extension of a soil or waste stack above the highest horizontal drain connected to the stack. (See Figure 1.2.53.)

Figure 1.2.53 Stack and stack terms.

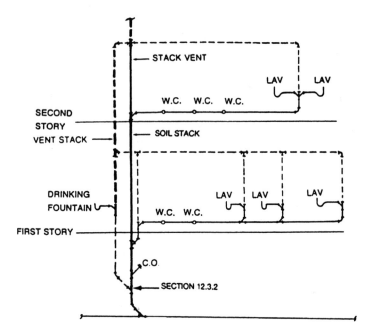

FIGURE 1.2.53 STACK AND STACK TERMS

Stack Venting

A method of venting a fixture or fixtures through the soil or waste stack. (See Figure 1.2.54.)

Figure 1.2.54 An example of stack venting.

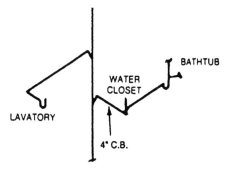

FIGURE 1.2.54 AN EXAMPLE OF STACK VENTING

Storm Sewer

A sewer used for conveying rain water, surface water, condensate, cooling water, or similar liquid wastes.

Subsoil Drain

A drain that collects subsurface or seepage water and conveys it to a place of disposal. (See Figure 1.2.55.)

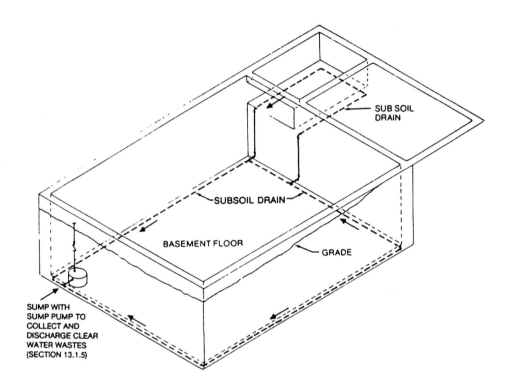

FIGURE 1.2.55 AN EXAMPLE OF SUBSOIL DRAIN

Figure 1.2.55   An example of a subsoil drain.

Sump

A tank or pit that receives liquid wastes only, located below the elevation of the gravity system and emptied by pumping. (See Figure 1.2.56.)

Figure 1.2.56 Sump.

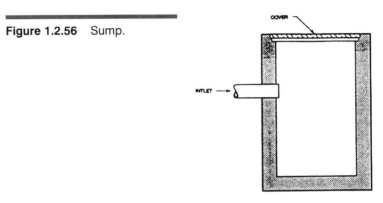

FIGURE 1.2.56 SUMP

Sump Pump

A permanently installed mechanical device other than an ejector for removing sewage or liquid waste from a sump. (See Figure 1.2.57.)

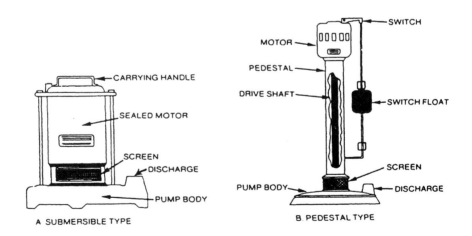

FIGURE 1.2.57 EXAMPLES OF SUMP PUMPS

Figure 1.2.57 Examples of sump pumps.

Supports

Devices for supporting and securing pipe, fixtures, and equipment. (See Figure 1.2.58.)

Figure 1.2.58 Typical pipe supports.

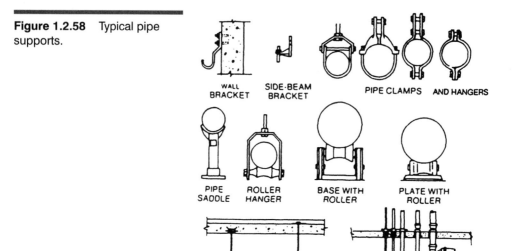

FIGURE 1.2.58 TYPICAL PIPE SUPPORTS

Tempered Water

Water at a temperature of not less than 90° and not more than 105°F.

Trap

A fitting or device that provides a liquid seal to prevent the emission of sewer gases without materially affecting the flow of sewage or waste water through it. (See Figure 1.2.59.)

Figure 1.2.59 A typical trap section view.

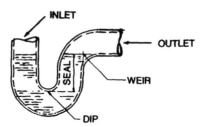

FIGURE 1.2.59 A TYPICAL TRAP SECTION VIEW

Trap Arm

A trap arm is that portion of a fixture drain between a trap and its vent. (See Figure 1.2.60.)

Figure 1.2.60 An example of a trap, trap arm, and trap primer arrangement.

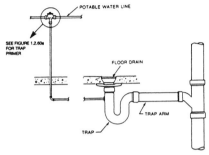

FIGURE 1.2.60b AN EXAMPLE OF TRAP, TRAP ARM, AND TRAP PRIMER ARRANGEMENT

Trap Primer

A trap primer is a device or system of piping to maintain a water seal in a trap. (See Figure 1.2.60.)

Figure 1.2.60 An example of a trap, trap arm, and trap primer arrangement.

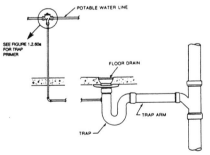

FIGURE 1.2.60b AN EXAMPLE OF TRAP, TRAP ARM, AND TRAP PRIMER ARRANGEMENT

Trap Seal

The maximum vertical depth of liquid that a trap will retain, measured between the crown weir and the top of the dip of the trap. (See Figure 1.2.59.)

Figure 1.2.59 A typical trap section view.

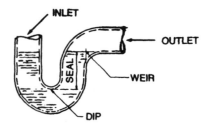

FIGURE 1.2.59 A TYPICAL TRAP SECTION VIEW

Vacuum

Any pressure less than that exerted by the atmosphere.

Vacuum Breaker, Nonpressure Type (Atmospheric)

A vacuum breaker that is not designed to be subject to static line pressure. (See Figure 1.2.61.)

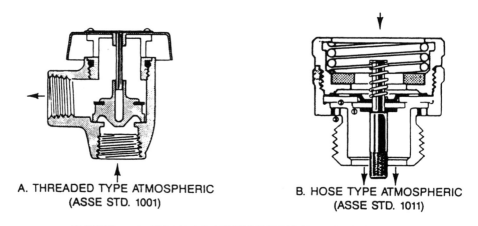

FIGURE 1.2.61 TYPES OF ATMOSPHERIC VACUUM BREAKERS

Figure 1.2.61 Types of atmospheric vacuum breakers.

Vacuum Breaker, Pressure Type

A vacuum breaker designed to operate under conditions of static line pressure. (See Figure 1.2.62.)

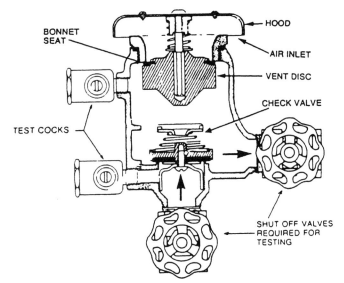

FIGURE 1.2.62 ONE TYPE OF PRESSURE TYPE VACUUM BREAKER (ASSE STD. 1020)

Figure 1.2.62 One type of pressure-type vacuum breaker (ASSE Std. 1020).

Vacuum Relief Valve

A device to prevent excessive vacuum in a pressure vessel. (See Figure 1.2.63.)

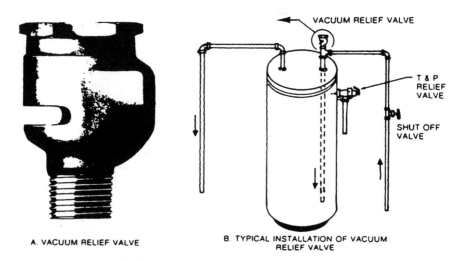

Figure 1.2.63 Vacuum relief valve.

Vent, Branch

A vent connecting one or more individual vents with a vent stack or stack vent. (See Figure 1.2.64.)

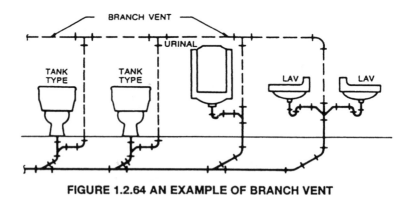

Figure 1.2.64 An example of a branch vent.

Vent, Circuit

A branch vent that serves two or more traps and extends from the downstream side of the highest fixture connection of a horizontal branch to the vent stack. (See Figure 1.2.65.)

Figure 1.2.65 An example of a circuit vent.

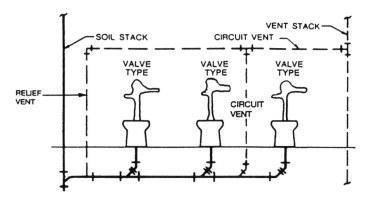

FIGURE 1.2.65 AN EXAMPLE OF A CIRCUIT VENT

Vent, Common

A vent connected at a common connection of two fixture drains and serving as a vent for both fixtures. (See Figure 1.2.66.)

Figure 1.2.66 An example of a common vent.

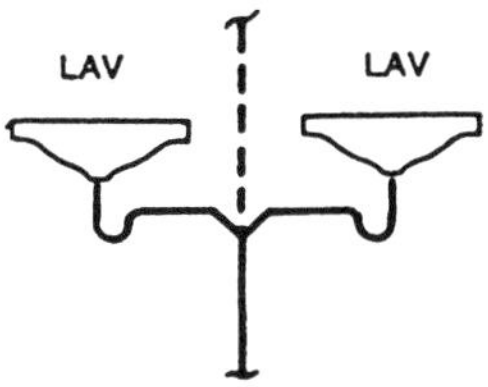

FIGURE 1.2.66 AN EXAMPLE OF COMMON VENT

Vent, Continuous

A vertical vent that is a continuation of the drain to which it connects. (See Figure 1.2.67.)

Figure 1.2.67 An example of a continuous vent.

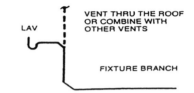

FIGURE 1.2.67 AN EXAMPLE OF CONTINUOUS VENT

Vent, Dry

A vent that does not receive the discharge of any sewage or waste.

Vent, Individual

A pipe installed to vent a fixture drain. It connects with the vent system above the fixture served or terminates outside the building into the open air. (See Figure 1.2.68.)

Figure 1.2.68 An example of an individual vent.

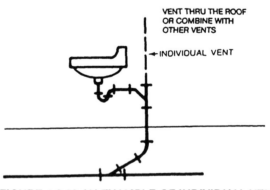

FIGURE 1.2.68 AN EXAMPLE OF INDIVIDUAL VENT

Vent, Loop

A circuit vent that loops back to connect with a stack vent instead of a vent stack. (See Figure 1.2.69.)

Figure 1.2.69 An example of a loop vent.

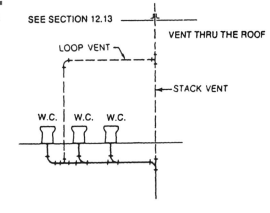

FIGURE 1.2.69 AN EXAMPLE OF LOOP VENT

Vent, Main

The principal artery of the venting system to which venting branches may be connected. (See Figure 1.2.70.)

Figure 1.2.70 An example of a main vent.

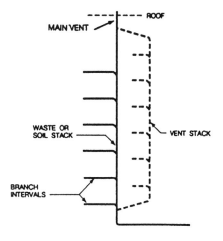

FIGURE 1.2.70 AN EXAMPLE OF MAIN VENT

Vent, Relief

An auxiliary vent that permits additional circulation of air in or between drainage and vent systems. (See Figure 1.2.71.)

Figure 1.2.71 Examples of relief vents.

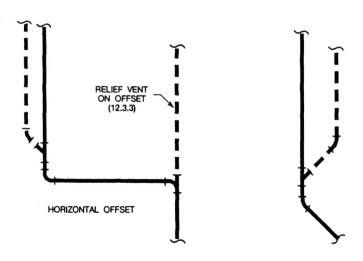

FIGURE 1.2.71 EXAMPLES OF RELIEF VENTS

Vent, Side

A vent connecting the drain pipe through a fitting at an angle not greater than 45° to the vertical. (See Figure 1.2.72.)

Figure 1.2.72 An example of a side vent.

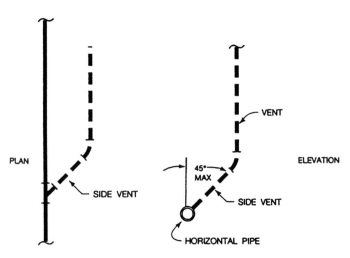

FIGURE 1.2.72 AN EXAMPLE OF SIDE VENT

Vent, Wet

A vent that receives the discharge of wastes other than from water closets and kitchen sinks. (See Figure 1.2.73.)

Figure 1.2.73 Examples of wet vent arrangements.

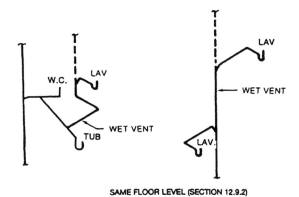

FIGURE 1.2.73 EXAMPLES OF WET VENT ARRANGEMENTS

Vent, Yoke

A pipe connecting upward from a soil or waste stack to a vent stack for the purpose of preventing pressure changes in the stack. (See Figure 1.2.74.)

Figure 1.2.74 Yoke vent arrangement.

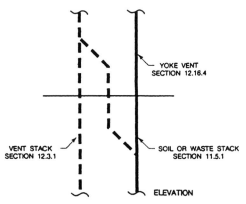

FIGURE 1.2.74 YOKE VENT ARRANGEMENT

Vent Pipe

Part of the vent system.

Vent Stack

A vertical vent pipe installed to provide circulation of air to and from the drainage system and which extends through one or more stories. (See Figure 1.2.53.)

Figure 1.2.53 Stack and stack terms.

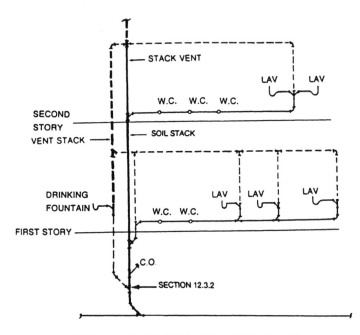

FIGURE 1.2.53 STACK AND STACK TERMS

Vent System

A pipe or pipes installed to provide a flow of air to or from a drainage system or to provide a circulation of air within such system to protect trap seals from siphonage and back pressure.

Waste Pipe

A pipe that conveys only waste.

Water Distributing Pipe

A pipe within the building or on the premises that conveys water from the water-service pipe to the point of usage. (See Figure 1.2.77.)

Figure 1.2.77 An example of a water-distributing pipe.

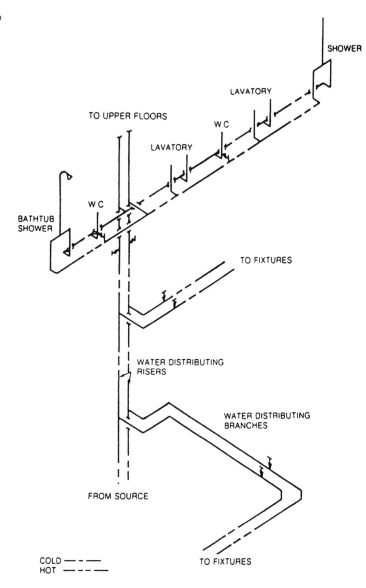

FIGURE 1.2.77 AN EXAMPLE OF WATER DISTRIBUTING PIPE

Water Service Pipe

The pipe from the water main, or other source of potable water supply, to the water-distributing system of the building served. (See Figure 1.2.78.)

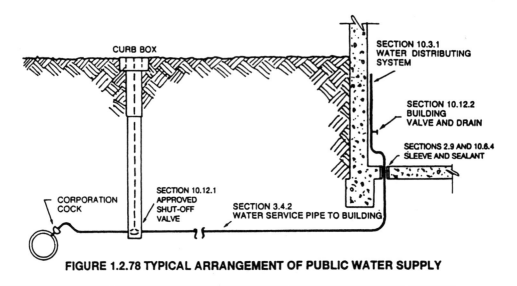

Figure 1.2.78 Typical arrangement of a public water supply.

Water Supply System

The water service pipe, the water-distributing pipes, and the necessary connecting pipes, fittings, control valves, and all appurtenances in or adjacent to the building or premises.

APPENDIX C

Tables and Charts

Type of Building Occupancy	Water Closets		Urinals		Lavatories		Bathtubs or Showers	Drinking Fountains	Other Fixtures
	No. of Persons	No. of Fixtures	No. of Persons	No. of Fixtures	No. of Persons	No. of Fixtures			
Assembly—places of worship									
Principal assembly place	300 men 150 women	1 1	300	1	1 per toilet room			1 for each 250 persons	
Educational and activities unit	250 men 125 women	1 1	250	1	1 per toilet room			1 for each 100 persons	
Assembly—other than places of worship, (auditoriums, convention halls, theaters)	1–100 101–200 201–400 Over 400, add 1 fixture for each additional 500 men; and 1 for each 300 women	1 2 3	1–200 201–400 401–600 Over 600, add 1 fixture for each 300 men	1 2 3	1–200 201–400 401–750 Over 750, add 1 fixture for each 500 persons	1 2 3		1 for each 500 persons	1 slop sink
Dormitories, school or labor, also institutional	Men: 1 for each 10 persons Women: 1 for each 8 persons		1 for each 25 men. Over 150, add 1 fixture for each 50 men		1 for each 12 persons (Separate dental lavatories should be provided in community toilet room. A ratio of 1 dental lavatory to each 50 persons is recommended.)		1 for each 8 persons; for women's dormitories, addition 1 bathtub should be installed at the ratio of 1 for each 30 women	1 for each 75 persons	Laundry trays 1 for each 50 persons. Slop sinks, 1 for each 100 persons
Single dwellings	1 for each dwelling unit		1 for each dwelling unit		1 for each dwelling unit				1 kitchen sink
Dwellings—multiple or apartment	1 for each dwelling unit or apartment				1 for each dwelling unit or apartment		1 for each dwelling unit or apartment		1 sink for each dwelling unit

Source: Table 4 ANSI A40–1993 Standard, Safety Requirements for Plumbing

Type of Building Occupancy	Water Closets		Urinals		Lavatories		Bathtubs or Showers	Drinking Fountains	Other Fixtures
	No. of Persons	No. of Fixtures	No. of Persons	No. of Fixtures	No. of Persons	No. of Fixtures			
Industrial—factories, warehouses, foundries, and similar establishments	1–10 11–25 26–50 51–75 76–100 1 fixture for each additional 30 employees	1 2 3 4 5	11–30 31–80 81–160 161–240 Over 240, add 1 for each 50 men	1 2 3 4	1–100 Over 100	1 to 10 persons 1 to 15 persons	1 shower for each 15 persons exposed to excessive heat or to occupational hazard from poisonous, infectious, or irritating material	1 for each 75 persons	
Institutional—other than hospitals or penal institutions (on each occupied story)	1 for each 25 men 1 for each 20 women		1 for each 50 men		1 for each 10 persons		1 for each 10 persons	1 for each 50 persons	
Hospitals—individual room, wards	1 1 for each 8 patients				1 1 for each 10 patients		1 1 for each 20 patients	1 for each 100 patients	Minimum 1 slop sink per floor for first 50 beds; 1 additional for each additional 50 or major fraction thereof
Hospitals—waiting rooms, employees	1 Same as public buildings		Same as public buildings		1 Same as public buildings			Same as public buildings	
Penal institutions—prisoners	1 in each cell 1 in each exercise room		1 in each exercise room		1 in each cell 1 in each exercise room			1 on each cell block floor 1 in each exercise room	One slop sink per floor

Source: Table 4 ANSI A40-1993 Standard, Safety Requirements for Plumbing

Type of Building Occupancy	Water Closets		Urinals		Lavatories		Bathtubs or Showers	Drinking Fountains	Other Fixtures
	No. of Persons	No. of Fixtures	No. of Persons	No. of Fixtures	No. of Persons	No. of Fixtures			
Penal institutions—employees	Same as public buildings		Same as public buildings		Same as public buildings			Same as public buildings	
Public buildings, offices, business mercantile, storage, and institutional—employees	1–15 16–35 36–55 56–80 81–110 111–150 1 fixture for each additional 40 employees	1 2 3 4 5 6	Urinals may be provided in men's toilet rooms in lieu of water closets but for not more than ½ of the required number of water closets		1–15 16–35 36–60 61–90 91–125 1 fixture for each additional 45 persons	1 2 3 4 5		1 for each 75 persons	1 slop sink per floor
Schools—elementary	30 boys/25 girls	1 each group	25	1	35 boys/girls	1 each group	In gym or pool shower rooms, 1/5 pupils of a class	1/40 pupils	
secondary	40 boys/30 girls	1 each group	25	1	40 boys/girls	1 each group		1/50 pupils	
Workers' temporary facilities	30 men 15 women	1 1	30	1	30 men 30 women	1 1		1 fixture or equivalent for each 100 workers	
Restaurants, pubs, and lounges	1–50 51–150 151–300 Over 300, add 1 fixture for each additional 200 persons	M F 1 1 2 2 3 4	1–150 Over 150 persons, add 1 fixture for each 150 men	1	1–150 151–200 200–400 Over 400, add 1 fixture for each additional person	1 2 3			

Source: Table 4 ANSI A40–1993 Standard, Safety Requirements for Plumbing

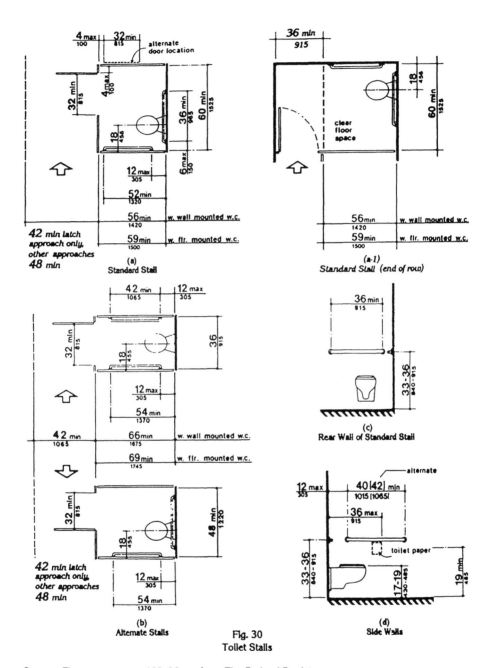

Fig. 30
Toilet Stalls

Source: Figures on pages 189–96 are from *The Federal Register.*

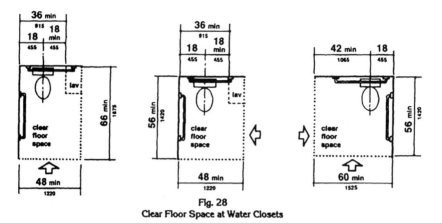

Fig. 28
Clear Floor Space at Water Closets

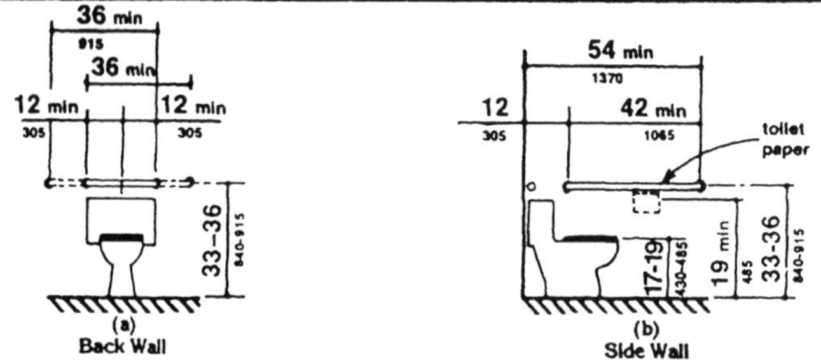

Fig. 29
Grab Bars at Water Closets

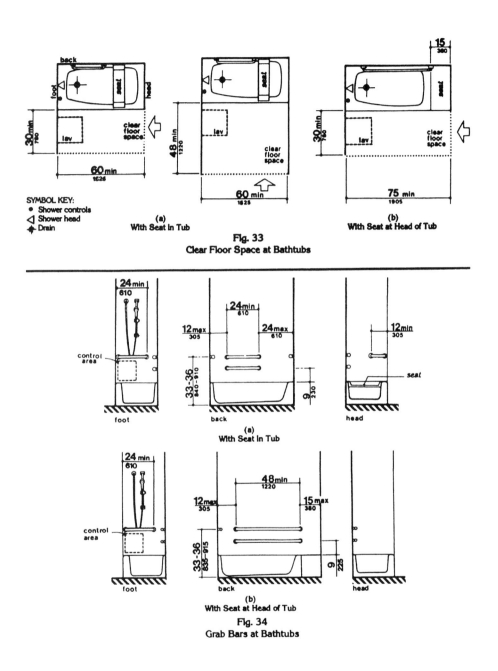

Fig. 33
Clear Floor Space at Bathtubs

Fig. 34
Grab Bars at Bathtubs

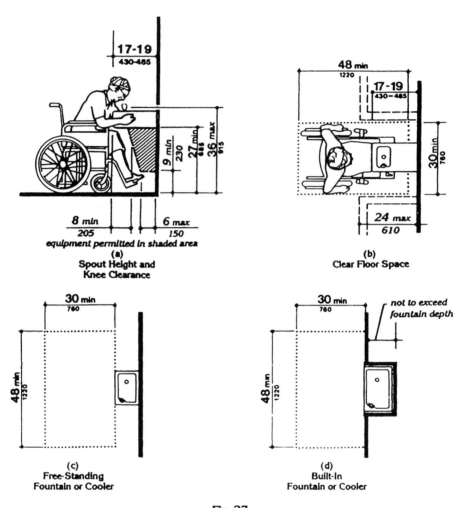

Fig. 27
Drinking Fountains and Water Coolers

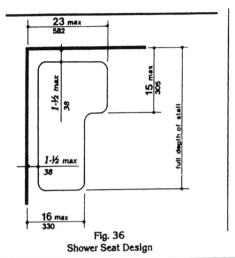

Fig. 36
Shower Seat Design

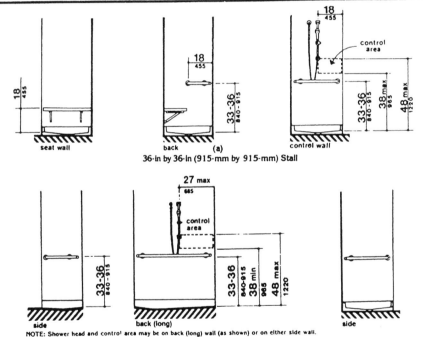

Fig. 37
Grab Bars at Shower Stalls

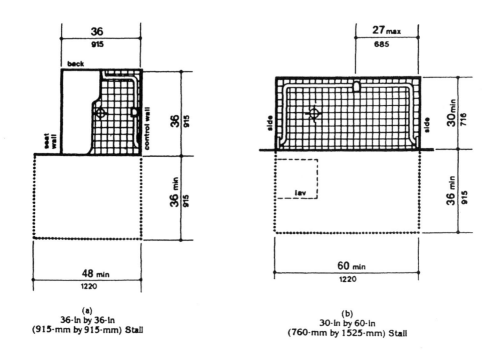

Fig. 35
Shower Size and Clearances

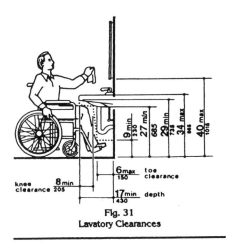

**Fig. 31
Lavatory Clearances**

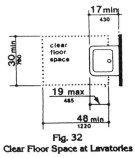

**Fig. 32
Clear Floor Space at Lavatories**

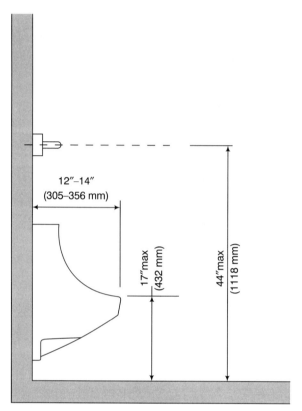

ADA mounting heights for urinals.

MINIMUM SPACE REQUIREMENTS FOR ENCLOSED PLUMBING FIXTURE SUPPORTS

This chart is for Architects, Plumbing Engineers, Contractors and others who specify enclosed wall supports for off-the-floor plumbing fixtures.

The specific purpose is to insure that specifiers provide sufficient wall space to properly install and conceal the types of supports shown.

All dimensions shown are minimum and can be increased to accommodate required variations or accessories.

The types of fixture supports shown are detailed in ANSI A112.6.1M.

PLUMBING AND DRAINAGE INSTITUTE

Reprinted with permission.

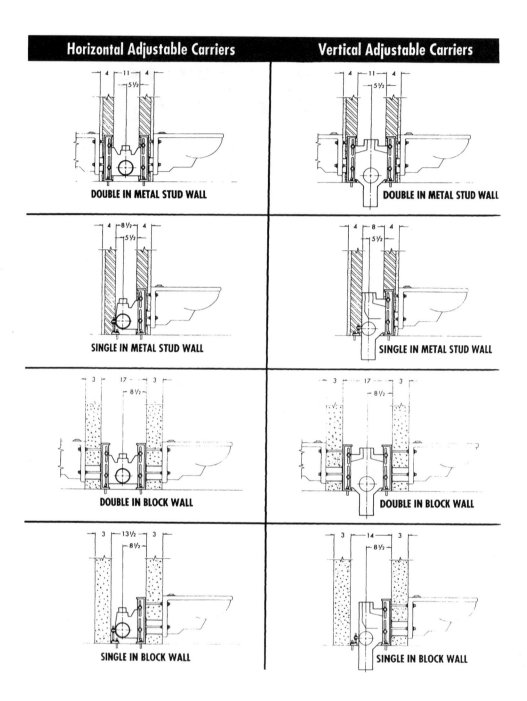

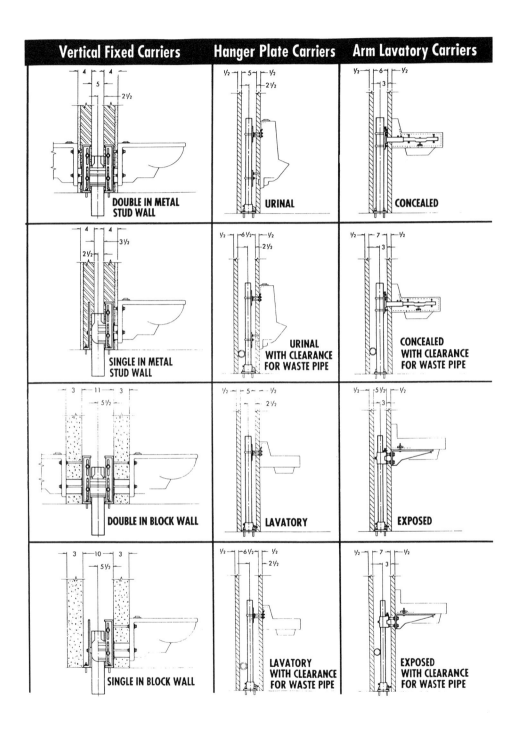

APPENDIX

Common Abbreviations, Symbols, and Conversion Factors for Plumbing Systems

ABBREVIATIONS

AD	Area drain
BFP	Backflow preventer
BS	Bar sink
CI	Cast iron
CO	Cleanout
CY	Cubic yard
DEG	Degree
DF	Drinking fountain
DHW	Domestic hot water
DIA	Diameter
DW	Dishwasher
DWV	Drain, waste, and vent
ELEV	Elevation
°F	Degrees Fahrenheit
FD	Floor drain
FH	Fire hydrant
FIX	Fixture
FPM	Feet per minute
FPS	Feet per second
FT	Feet

GAL	Gallon
GALV	Galvanized
GPM	Gallons per minute
HB	Hose bibb
HP	Horsepower
HRS	Hours
HU	Hub up
HV	Hose valve
HW	Hot water
HWC	Hot water circulation
HWR	Hot water return
ID	Inside diameter
IN	Inches
INV	Invert
KIT	Kitchen
KS	Kitchen sink
LAV	Lavatory
LDR	Leader
LF	Lineal foot
M	Meter
MAX	Maximum
MB	Mop basin
MH	Manhole
MIN	Minimum
OD	Outside diameter
OS & Y	Outside screw and yoke valve
PIV	Post indicator valve
PRV	Pressure-reducing valve

Common Abbreviations, Symbols, and Conversion Factors

PSI	Pounds per square inch
P & T	Pressure and temperature relief valve
PVC	Polyvinyl chloride plastic
[R]	Rough-in only
RCP	Reinforced concrete pipe
RD	Roof drain
RED	Reducer
RL	Roof drain leader
S	Sink
SA	Shock Absorber
SAN	Sanitary
SC	Sill cock
SH	Shower
SF	Square foot
SQ.IN.	Square inch
SS	Service sink
SV	Service weight
UR	Urinal
V	Vent
VB	Vacuum breaker
VC	Vitrified clay
VTR	Vent through roof
WC	Water closet
WCO	Wall cleanout
WH	Wall hydrant
WHA	Water hammer arrestor
XH	Extra heavy

MECHANICAL SYMBOLS

Piping Symbols

————— AR —————   Acid waste piping

————— A —————   Compressed air

——————————   Soil or waste piping above floor

▬▬▬▬▬▬▬▬   Soil or waste piping below floor

— — — — — — —   Vent piping

————— —— —————   Cold water piping

———— — — ————   Hot water piping

——— —— — ———   Hot water recirculation piping

——— — — 140° ———   140°F hot water piping

——— — — 180° ———   180°F hot water piping

————— G —————   Gas piping

————— R —————   Rainwater piping

————— H —————   Humidity piping

————— F —————   Fire sprinkler piping

———————O   Elbow up

———————⊖   Elbow down

————O————   Tee up

————⊖————   Tee down

Valve and Fitting Symbols

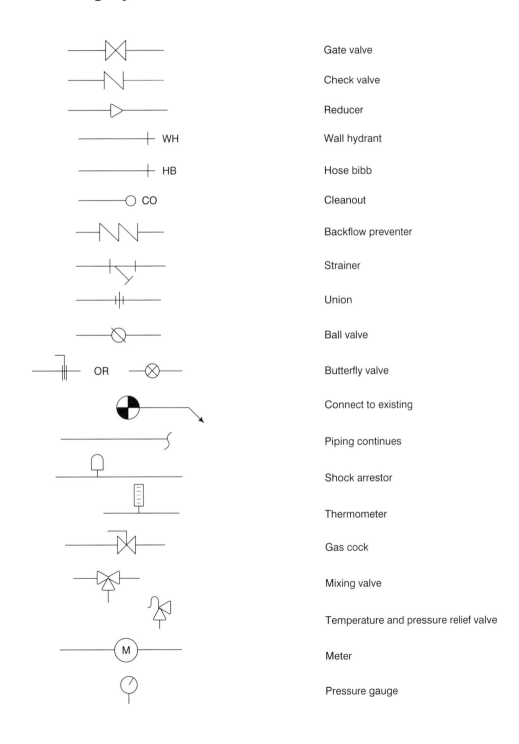

Common Abbreviations, Symbols, and Conversion Factors

Fixture Symbols

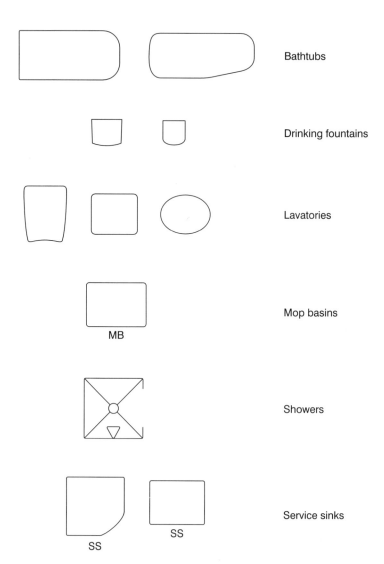

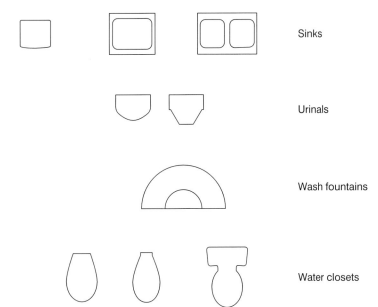

Conversion Factors

MULTIPLY	BY	TO OBTAIN
Acres	43.560	Square feet
Acres	4047	Square meters
Acres	1.562×10^{-3}	Square miles
Acre-feet	43,560	Cubic feet
Acre-feet	3.259×10^{5}	Gallons
Atmospheres	76.0	Centimeters of mercury
Atmospheres	29.92	Inches of mercury
Atmospheres	33.90	Feet of water
Atmospheres	10.333	Kgs per square meter
Atmospheres	14.70	Pounds per square inch
Centimeters	0.3937	Inches
Centimeters of mercury	0.01316	Atmospheres
Centimeters of mercury	0.4461	Feet of water
Centimeters of mercury	27.85	Pounds per square foot
Cubic centimeters	3.531×10^{-3}	Cubic feet
Cubic centimeters	6.102×10^{-2}	Cubic inches
Cubic centimeters	1.308×10^{-6}	Cubic yards
Cubic feet	1728	Cubic inches
Cubic feet	0.02932	Cubic meters
Cubic feet	0.03704	Cubic yards
Cubic feet	7.481	Gallons
Cubic feet	28.32	Liters
Cubic feet per minute	0.1247	Gallons per second
Cubic feet per minute	0.4720	Liters per second
Cubic feet per minute	62.4	Lbs of water per minute
Cubic inches	16.39	Cubic centimeters
Cubic inches	5.787×10^{-1}	Cubic feet
Cubic inches	1.639×10^{-5}	Cubic meters
Cubic inches	2.143×10^{-5}	Cubic yards
Cubic inches	4.329×10^{-3}	Gallons
Cubic inches	1.639×10^{-2}	Liters
Cubic meters	35.31	Cubic feet
Cubic meters	61,023	Cubic inches
Cubic meters	1.308	Cubic yards
Cubic meters	264.2	Gallons
Cubic yards	7.646×10^{-5}	Cubic centimeters
Cubic yards	27	Cubic feet
Cubic yards	46,656	Cubic inches
Cubic yards	202.0	Gallons
Cubic yards per minute	0.45	Cubic feet per second
Cubic yards per minute	3.367	Gallons per second

Cubic yards per minute	12.74	Liters per second
Feet	30.48	Centimeters
Feet	12	Inches
Feet	0.3048	Meters
Feet	1/3	Yards
Feet of water	0.02050	Atmospheres
Feet of water	0.8826	Inches of mercury
Feet of water	304.8	Kgs per square meter
Feet of water	62.43	Pounds per square feet
Feet of water	0.4335	Pounds per square inch
Feet per minute	0.5080	Centimeters per second
Feet per minute	0.01667	Feet per second
Feet per minute	0.01829	Kilometers per hour
Feet per minute	0.3048	Meters per minute
Feet per minute	0.01136	Miles per hour
Feet per second	30.48	Centimeters per second
Feet per second	1.097	Kilometers per hour
Feet per second	0.5921	Knots per hour
Feet per second	18.29	Meters per minute
Feet per second	0.6818	Miles per hour
Feet per second	0.01136	Miles per minute
Gallons	3785	Cubic centimeters
Gallons	0.1337	Cubic feet
Gallons	231	Cubic inches
Gallons	3.785×10^{-3}	Cubic meters
Gallons	4.951×10^{-3}	Cubic yards
Gallons	3.785	Liters
Gallons	8	Pints (liquid)
Gallons	4	Quarts (liquid)
Gallons per minute	$2,228 \times 10^{-3}$	Cubic feet per second
Gallons per minute	0.06308	Liters per second
Inches	2.540	Centimeters
Inches	10	Mils
Inches of mercury	0.03342	Atmosphere
Inches of mercury	1.133	Feet of water
Inches of mercury	345.3	Kgs per square meter
Inches of mercury	70.73	Pounds per square feet
Inches of mercury	0.4912	Pounds per square inch
Inches of water	0.002458	Atmospheres
Inches of water	0.07355	Inches of mercury
Inches of water	5.204	Pounds per square feet
Inches of water	0.03613	Pounds per square inch
Kilograms	2.2045	Pounds
Kilograms	1.102×10^{-3}	Tons (short)

Kgs per cubic meter	0.06243	Pounds per cubic foot
Kgs per cubic meter	3.613×10^{-5}	Pounds per cubic inch
Kgs per meter	0.6720	Pounds per foot
Kgs per square meter	9.678×10^{-5}	Atmospheres
Kgs per square meter	3.281×10^{-2}	Feet of water
Kgs per square meter	2.896×10^{-3}	Inches of mercury
Kgs per square meter	0.2048	Pounds per square foot
Kilometers	3281	Feet
Kilometers	0.6214	Miles
Kilometers	1093.6	Yards
Kilometers per hour	54.68	Feet per minute
Kilometers per hour	0.9113	Feet per second
Kilometers per hour	0.6214	Miles per hour
Liters	10^3	Cubic centimeters
Liters	0.03531	Cubic feet
Liters	61.02	Cubic inches
Liters	10^{-3}	Cubic meters
Liters	1.308×10^{-3}	Cubic yards
Liters	0.2642	Gallons
Liters	2.113	Pints (liquid)
Liters	1.057	Quarts (liquid)
Liters per minute	5.855×10^{-6}	Cubic feet per second
Liters per minute	4.403×10^{-3}	Gallons per second
Meters	100	Centimeters
Meters	3,208	Feet
Meters	39.37	Inches
Meters	10^{-3}	Kilometers
Meters	10^3	Millimeters
Meters per minute	1.667	Centimeters per second
Meters per minute	3.281	Feet per minute
Meters per minute	0.05468	Feet per second
Meters per minute	0.06	Kilometers per hour
Meters per minute	0.03728	Miles per hour
Meters per second	1968	Feet per minute
Meters per second	3.284	Feet per second
Meters per second	3.0	Kilometers per hour
Meters per second	0.06	Kilometers per minute
Meters per second	2.237	Miles per hour
Miles	1.609×10^6	Centimeters
Miles	1.6093	Kilometers
Miles per hour	44.70	Centimeters per second
Miles per hour	88	Feet per minute
Miles per hour	1.467	Feet per second
Miles per hour	1.6093	Kilometers per hour
Miles per hour	26.82	Meters per minute

Millimeters	0.03937	Inches
Pints (dry)	33.60	Cubic inches
Pints (liquid)	28.87	Cubic inches
Pounds	7000	Grains
Pounds	453.6	Grams
Pounds	16	Ounces
Pounds of water	0.01602	Cubic feet
Pounds of water	27.68	Cubic inches
Pounds of water	0.1198	Gallons
Pounds of water per minute	2669×10^{-6}	Cubic feet per second
Pounds per cubic foot	0.01602	Grams per cubic centimeter
Pounds per cubic foot	16.02	Kgs per cubic meter
Pounds per cubic inch	27.68	Pounds per cubic centimeter
Pounds per foot	1.488	Kgs per meter
Pounds per inch	178.6	Grams per centimeter
Pounds per square foot	0.01602	Feet of water
Pounds per square foot	4.882	Kgs per square meter
Pounds per square foot	6.944×10^{-3}	Pounds per square inch
Pounds per square inch	0.06804	Atmospheres
Pounds per square inch	2.307	Feet of water
Pounds per square inch	2.036	Inches of mercury
Pounds per square inch	703.1	Kgs per square meter
Pounds per square inch	144	Pounds per square foot
Quarts (dry)	67.20	Cubic inches
Quarts (liquid)	57.75	Cubic inches
Square centimeters	1.076×10^{-3}	Square feet
Square centimeters	0.1550	Square inches
Square centimeters	10^{-6}	Square meters
Square centimeters	100	Square millimeters
Square feet	2.296×10^{-5}	Acres
Square feet	929.0	Square centimeters
Square feet	144	Square inches
Square feet	0.09290	Square meters
Square feet	3.587×10^{-3}	Square miles
Square feet	0.1296	Square varas
Square feet	1/9	Square yards
Square inches	6.452	Square centimeters
Square inches	6.944×10^{-3}	Square feet
Square inches	10^4	Square mils
Square inches	645.2	Square millimeters
Square kilometers	247.1	Acres
Square kilometers	10.76×10^4	Square feet
Square kilometers	0.3861	Square miles
Square kilometers	1.196×10^3	Square yards
Square meters	2.471×10^{-1}	Acres

Square meters	10.764	Square feet
Square meters	3.861×10^{-7}	Square miles
Square meters	1.196	Square yards
Square miles	640	Acres
Square miles	27.88×10^{5}	Square feet
Square miles	2.590	Square kilometers
Square millimeters	1.550×10^{-3}	Square inches
Square yards	2.066×10^{-4}	Acres
Square yards	0.8361	Square meters
Square yards	3.228×10^{-7}	Square miles
Temp (°C) +17.8	1.8	Temp (°F)
Temp (°F) -32	5/9	Temp (°C)
Tons (long)	1016	Kilograms
Tons (long)	2240	Pounds
Tons (metric)	10^{3}	Kilograms
Tons (metric)	2205	Pounds
Tons (short)	907.2	Kilograms
Tons (short)	2000	Pounds
Yards	91.44	Centimeters
Yards	0.9144	Meters

Index

A. O. Smith Water Products, Accu-U-Size, 130
Acrylonitrile Butadiene Styrene (ABS), 25
Aeration units, 41, 44
Air chambers, 95, 97
Air gaps, 101
Americans with Disabilities Act (ADA), 56, 62
ANSI-A40 Standard, 1, 16
Asbestos-Cement piping, 31

Backflow preventers, 101
   Application table 8-6, 103
Bathtubs, 54
Brass, pipe and fittings, 28–29
Building drain
   Sizing, 69

Carbon steel piping and fittings, 26
   Ductile iron, 27
   Joints, 27
   Schedules, 27
   Stainless steel, 27

Cast iron pipe and fittings, 25
   Extra heavy, 25
   Joints, 26
   Service weight, 25
Chairs, closet, 51–52
Chlorinated Polyvinyl Chloride (CPVC), 25
Circuit vents, 18, 74
Clay tile pipe, 31
Cleanouts, 57
Combustion air, 126
Concrete pipe (RCP), 31
Copper pipe and fittings, 29
   Color coding, 29
   Expansion and contraction, 105
   Joints, 29
   Types, 29

Diversity, 107
Domestic water systems, 2, 93
   Chemical composition, 100
   Heating, 124–131
   Noise, 95
   Pressure losses, 93, 94

Domestic water systems, *continued*
   Pressures, 94
   Purpose, 93
   Sizing, 107
   Temperature, 97
   Velocity, 95, 97
Drain, waste and vent (DWV), 2, 3
   Change of direction, 32
   Combination systems, 80
   Fittings, 32
   Sanitary systems, 2
   Sizing, 65
   Storm systems, 2
   Supports and hangers, 35
Drainage Fixture Unit (DFU), 65
   Values, table, 8
Drains, 56
   Area, 57, 80
   Floor, 56
   Roof, 57, 80–81
Drain and waste systems, 69–70
   Rules, 70
   Sizing, tables, 70–71
Drinking fountain, 56
Ductile iron, pipe and fittings, 27
Duriron pipe and fittings, 26

Electric coil, immersion type, 126
Electrolysis, 99
Expansion joints, 106

Fiberglass pipe and fittings, 25
Fixtures, plumbing, 7, 17, 49
   Bathtubs and showers, 54
   Cleanouts, 56
   Drinking fountains, 56
   Flood level rim, 17
   Floor drains, 7, 12, 56
   Layout, 49
   Location, 60
   Quantities, 60
   Roof drains, 56
   Sinks, 53–54
   Traps, 6, 7
   Types, 49
   Urinals, 52
   Water closets, 50
Flood level rim, 17

Floor drains, 7, 12, 56
Flush valves, 51

Glass pipe and fittings, 32
Grade (or slope), 3, 5, 94
   Sanitary
      Clearance, 5
      Minimums, 5
      Tables, 4
      Velocity, 3
   Vents
      Minimums, 16
Gutters and downspouts, 80

Hangers and supports, 35
Hot water recirculation systems, 99
Hunter curves, 66, 108–109
Hydrostatic test, 23, 26

Infrared sensing flush valves, 52
Instantaneous water heaters, 127–128
Interceptors 7
   Grease, 7
   Solids, 7

Joints
   Brazed, 29
   Cast iron, 26
   Mechanical joint, 27–28
   Neoprene gasket, 24, 26–27
   No-hub gasket, 24, 26–27
   Slip joint, 29–30
   Soldered, 29–30
   Steel, 27
   Victaulic, 27

Lavatories, 53
Layout, plumbing fixtures, 60–62
Lead, in drinking water, 29, 100
Lead pipe and fittings, 28
Legionella bacterium (Legionnaire's Disease), 99

Manning chart, 91
Materials, 21
   Selection criteria, 21–22
      Availability, 22
      Corrosion, 21

Costs, 22
Noise, 22
Pressure, 21
Temperature, 21
Mixing valves, 129

Nonferrous metals (See also copper pipe and fittings)
Aluminum, 28
Brass, 28–29
Copper, 28–30
Lead, 28

Overflow systems (storm), 86–87

"P" trap (see Trap)
Plastics, 22
Acrylonitrile Butadiene Styrene (ABS), 25
Advantages, 23
Chlorinated Polyvinyl Chloride (CPVC), 25
Disadvantages, 23
Joints, 23–24
Polybutylene (PB), 25
Polyethylene (PE), 25
Polypropylene, (PP), 25
Polyvinyl Chloride (PVC), 25
Polyvinylidene Fluoride (PVDF), 25
Ratings, 23
Plumbing and Drainage Institute (PDI), 52
Pumps, 57
Booster, 59, 94, 121
Hot water recirculation, 99
Sewage ejector, 58, 69
Storm water, 84
Sump, 58, 80

Quicky vent, 18–19

Rainfall rates, map, 84
Recovery, domestic hot water, 124–127
Relief valve, pressure and temperature, 127
Riser diagram, 67, 107
Roof drains, 80, 81
Roof drain systems (see stormwater systems)
Run-off coefficient, 89–90

Scuppers, 86
Septic systems, 41–42

Service sinks, 53–54
Sewage ejectors (See pumps)
Sewer gas, 6
Showers, 54–55
Sinks, 53
Slope (or grade), 3, 5, 69, 83
Clearance, 5
Minimum, 5
Table, 4
Velocity, 3
Soil percolation test, 44
Stabilization ponds, 41, 45
Stainless steel, 27
Standard Dimensional Ratio (SDR), 23
Stone-based pipe and fittings, 31
Asbestos-cement piping, 31
Clay pipe, 31
Concrete pipe, 31
Glass pipe and fittings, 32
Storage, domestic hot water, 124–127
Stormwater systems, 79–91
Combination systems, 80
Flow control systems, 81
Goals, 79
Gutter systems, 80
Overflow systems, 86
Site drainage systems, 79, 84–91
Sizing, 83–89
Horizontal branch sizing, Table 7-2, 86
Roof gutter sizing, Table 7-3, 87
Vertical conductor sizing, Table 7-1, 85
Supports and hangers, 35
Aboveground, 35
Fixture, 52
Spacing, 36
Underground, 35

Temperature and pressure relief valves, 127
Tests
Hydrostatic, 23, 26
Percolation, 44
Toilet (See water closet)
Total Developed Length (TDL), 72–75
Traps, 6
Arm, Table 2-3, 16
Bell, 7
Blowout, 15
Deep seal, 8

Traps, *continued*
  Dip, 7
  Drum, 7
  Evaporation, 7
  Integral, 7
  "P", 6
  Primer, 8
  "S", 7
  Seal, 7, 14
  Siphonage, 15
  Size, minimum, 7
  Sizing, 15
  Weir, 7

Urinal, 52

Valves, 37
  Ball, 38
  Butterfly, 38
  Check, 39
  Gate, 37
Velocity
  Sanitary, 3–5
  Water, 95–97
Venting, 12–14
  Automatic (Quicky), 18
  Branch vents, 73
  Branch vent sizing, Table 6-4, 75
  Circuit, 18, 74
  Combination waste and vent systems, 18
  Frost closure, 14
  Goals, 14
  Loop and circuit venting sizing, Table 6-3, 74
  Loop vents, 18, 74
  Sizing, 73
  Terminations, 14
  Wet vents, 17–18
Victaulic joints, 27

Waste systems, 18
Water closets, 50
  Chairs, 51
  Types, 50
Water hammer, 95
Water heaters, 125–129
  Fundamentals, 124
  Sizing, 130
  Types
    Instantaneous, 127
    Tank types, 126
Water meters, 107, 122
Water softener, 122
Water Supply Fixture Units (WSFU), 66, 107
Water supply systems
  Private, 47
  Public, 46
Wet wall, 62

TH
6123
.W46
1997

40.00